Biology

Revision Notes

Author
Marilyn Brodie

Series editor
Alan Brewerton

A level

EDUCATIONAL

Every effort has been made to trace copyright holders and to obtain their permission for the use of copyright material. The authors and publishers will gladly receive information enabling them to rectify any error or omission in subsequent editions.

First published 1998

Letts Educational, Schools and Colleges Division, 9–15 Aldine Street, London W12 8AW
Tel. 0181 740 2270
Fax 0181 740 2280

Text © Marilyn Brodie 1998

Editorial, design and production by Hart McLeod, Cambridge

British Library Cataloguing-in-Publication Data

A CIP record for this book is available from the British Library

ISBN 1 84085 096 5

Printed and bound in Great Britain

Letts Educational is the trading name of BPP (Letts Educational) Ltd

Contents

Introduction

This book is designed to help you with your revision for A level biology. It is organised in such a way that it is appropriate regardless of whether you are following a modular scheme or studying on a course which involves final examinations.

It is not intended to be a substitute for your own notes or textbooks. Rather it should be viewed as an additional part of your revision material.

You will notice a number of annotations and tips have been written in boxes throughout the text. These are designed to help you learn and give you some hints about revision techniques. Plenty of further space has been made available so that you can add your own notes such as references to notes or textbooks. Many of the 'standard' biological diagrams are missing – it is suggested that you refer to your textbook or notes for these and where appropriate, practice drawing and/or labelling them.

The book is not allied to any one syllabus but is designed to cover all the main points found in all the major examination boards. It is important that you become familiar with your own particular syllabus so that you know if a particular area is needed and how much detail is required.

This book is not a biology text book but is meant to be used as a guide to revision. If you plan your revision around its contents and chapters then you will have covered all the main areas on which you are likely to be asked questions. There are a few hints that will help you remain in control of your revision.

Organisation is the key:

- Organise your notes into an order that matches your syllabus (you may find the chapter headings in this book helpful with this)

- Organise your time
  - draw up a revision timetable so that you can see how much time you have before each of your examinations
  - start your revision a long time before your examination because that way your revision sessions need only be short. If you try to do too much, you will lose concentration and become frustrated.

- Devise a personal revision file in the form of one page summaries, overviews, spider diagrams, index cards, etc.

- The process of deciding the most important material for a summary will concentrate your mind and you will be learning as you go along.

- It is these summaries, etc. you can look through quickly the evening before your examination: you won't have time to read either your notes or textbook.

There is no substitute for good, thorough revision and that cannot start soon enough.

If you feel confident that you have revised thoroughly, you will go into your examinations with the same confidence.

Good luck!

Chapter one

The processes of life

Glossary

ATP/ADP/AMP – adenosine **tri** phosphate (3 phosphate groups) / adenosine **di** phosphate (2 phosphate groups) / adenosine **mono** phosphate (1 phosphate group)

Autotrophic – literally, **auto** – self, **trophic** – feeding, – can make its own food

Biological scales – units to remember:

$1m = 1000mm$ (millimetres) $= 10^{-3}m$

$1mm = 1000\mu m$ (micrometres) $= 10^{-6}m$

$1\mu m = 1000nm$ (nanometres) $= 10^{-9}m$

Condensation – reaction in which water is removed

Decarboxylation – reaction in which carbon dioxide is removed

Enzymes – globular proteins which catalyse specific biochemical reactions

Eukaryotic – literally, **eu** – after, **karyote** – nucleus

Glycolysis – literally, **glyco** – glucose, **lysis** – splitting

Hydrolysis – reaction in which water is added to **split** a compound.

Limiting factors – when a chemical process depends on more than one essential condition being favourable, its rate is limited by that factor nearest its minimum value

Macromolecule – a giant biological molecule made up of repeating units, sometimes called a polymer

Oxidation – direct addition of oxygen, removal of hydrogen atoms or removal of electrons

Phosphorylation – reaction in which phosphate is added

Photosynthesis – literally, **photo** – light, **synthesis** – to make

Polymer – a long chain molecule consisting of repeated subunits, e.g. a polysaccharide

Prokaryotic – literally, **pro** – before, **karyote** – nucleus

Proteins

– **primary structure** – linear sequence of amino acids

– **secondary structure** – the folding of a polypeptide chain due to hydrogen bonding; there are two main types, α – helix and β – pleated sheet

– **tertiary structure** – folding and coiling of polypeptide chain producing a 3-dimensional molecular shape

– **quaternary structure** – proteins possessing more than one type of polypeptide chain

Reduction – removal of oxygen, addition of hydrogen atoms or addition of electrons

> Hydro = water
> lysis = splitting

> Make sure that you understand the difference between atoms, molecules, organisms and cells. Examiners state that a large number of candidates still confuse them.

The chemicals of life

The most important elements in living organisms are **carbon, hydrogen, oxygen, nitrogen, calcium, sulphur** and **phosphorus**.

Life on earth, in all its forms, is based on the unique chemical properties of **carbon** which can form strong chemical bonds both with itself and with hydrogen.

The four main groups of macromolecules found in living organisms are proteins, carbohydrates, lipids (fats) and nucleic acids (there are other smaller molecules and ions).

Characteristics of the important molecules found in living organisms

Inorganic compounds

Compounds	Notes
Ions in solution in water*, e.g. H (hydrogen), Na^+ (sodium), Ca^{2+} (calcium)	Ions have a very varied role, e.g. Ca^{2+} in muscle contraction and synaptic transmission
Carbon dioxide	In solution as HCO_3^- ions (hydrogen carbonate)
Insoluble compounds	Bone is made up of 70% salts especially calcium phosphate

* The importance of **water** as a medium for life comes from its properties as a **solvent**, its **heat capacity**, **surface tension** and **freezing** and **evaporation properties**.

Organic compounds

Basic structure of biological molecules

Molecule	Polymer
Glucose	
Amino acid	
Lipid	

Lipids can be hard to remember. Glycerol has 3 carbon atoms with an OH on one side of each and H everywhere else. If you write fats as HOOC.R and remember to remove one H_2O for each fatty acid, you'll remember how to make a lipid.

If you are not studying chemistry and find chemical formulae hard – learn the basic structure of glucose and an amino acid and that you can join them by removing $H_2O \rightarrow$ carbohydrates and proteins.

Compounds	Notes
Biological polymers are long-chain molecules made up of smaller molecules (monomers) connected by chemical bonds include **polysaccharides**, **proteins** and **lipids**	Polysaccharides – structure and storage Proteins – growth and repair Lipids – storage and cell metabolism Nucleic acids – protein synthesis and inheritance
The chemical reaction that joins monomers is **condensation**, **hydrolysis** is the splitting of a polymer into monomers	Each reaction has its own specific enzyme
	Starch, glycogen and cellulose are examples of polysaccharides Glucose and fructose are reducing sugars, i.e. produce a red precipitate when tested with Benedict's solution Starch gives blue/black colour with iodine solution 20 common amino acids combine in many ways to make proteins

Compounds	Notes
	Biuret solution tests positive for protein (pale mauve) Includes fats and oils
	Phospholipid important in membrane structure (P replaces one fatty acid)
	Fatty acids can be **saturated** (no double bonds) or **unsaturated** (double bonds in the hydrocarbon chain)

Distinguishing between some important biological polymers:

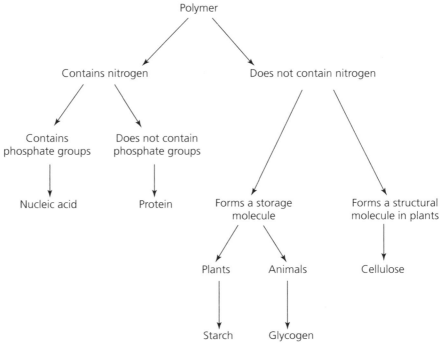

Enzymes

Enzymes are biological catalysts which speed up biochemical reactions. They are tertiary proteins and their molecular shape results in them being **specific** in their action.

An enzyme needs to form a precise fit with the molecule or molecules taking part in the reaction (known as the **substrate**). The part of the enzyme which makes contact with the substrate is called the **active site**.

The substrate (S) binds to the active site of the enzyme (E) forming an **enzyme-substrate complex** (ES), the reaction takes place and the **products** (P) are released with the enzyme unchanged and ready to catalyse another reaction.

$$E + S \rightleftharpoons ES \rightleftharpoons E + P$$

Lock and key hypothesis of enzyme activity

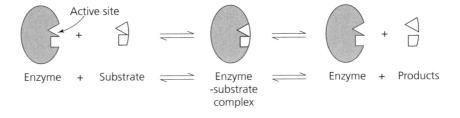

Factors which affect enzyme-catalysed reactions

1 Temperature

A rise in temperature affects the rate of reaction by increasing the speed of the molecules, so they are more likely to collide. The rate of reaction is at its greatest when the most collisions occur. This is known as the **optimum temperature**. However, enzymes are proteins and their tertiary structure is held together by ionic, covalent and hydrogen bonds,which become unstable as temperature rises. Above 40°C bonds will break and the structure is changed, including the shape of the active site. This is called **denaturation**, which above 45–50°C is often permanent with an irreversible change in structure.

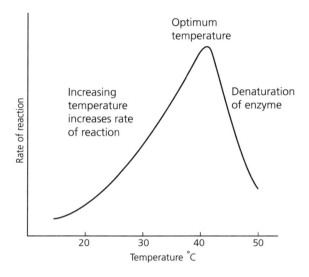

2 pH

Most enzymes are active over a narrow pH range either side of the pH at which the rate of reaction is at its maximum – the optimum pH. Extremes result in denaturation; the pH alters the charges of acidic and basic groups and the active site can no longer function. This denaturation may be temporary or permanent.

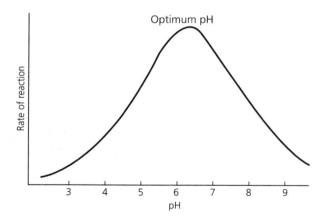

3 Substrate concentration

For a given amount of enzyme, increasing the concentration of substrate will initially increase the rate of reaction. The rate then levels out as all the enzyme molecules are taken up by substrate molecules. This means that the rate is now determined by the length of time the enzyme and substrate are combined. The rate of this reaction will only increase now if more enzyme is added.

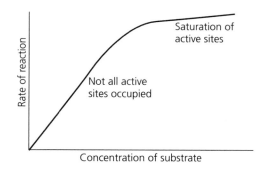

4 Enzyme concentration

If the substrate concentration is at a constantly high level and temperature and pH remain constant, the rate of reaction is directly proportional to the concentration of enzyme.

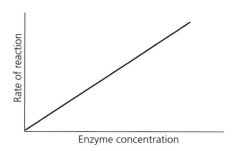

Enzyme inhibition

Inhibitors are substances which reduce enzyme activity.

There are a number of types of inhibitors:

1 Competitive inhibitors

This type of inhibitor has a similar structure to the normal substrate molecule and competes for the active site. It reduces the rate because it, instead of the substrate, occupies the enzyme's active site. If more substrate is added the effect of the inhibitor is reduced.

2 Non-competitive inhibitors

This type of inhibitor is not similar in structure to the substrate and does not combine at the active site of the enzyme. It combines elswhere and alters the globular shape of the enzyme so that either the substrate cannot combine, or does combine but no products are formed. Adding more substrate will not reduce the degree of inhibition as this is dependent on the amount of inhibitor present.

Substrates compete for active site

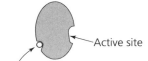

Active site

Non-competitive inhibitor joins at another site and alters the active site shape.

Structure and function of cells

Cells were first described in 1665 by Robert Hooke.

The **cell theory** states that the cell is the basic unit of an organism.

Cells can be observed with various kinds of microscope, e.g. a typical **light microscope** magnifies about 800 times, while an **electron microscope** can magnify objects 300,000 times.

There are two basic types of cell: **prokaryotic** cells and **eukaryotic** cells.

Prokaryotes such as bacteria are usually single-celled organisms.

Eukaryotes can be as simple as yeast or as complex as humans.

Comparing prokaryotes and eukaryotes

Prokaryotes	Eukaryotes
Average diameter of cells 0.5–5µm	Cells usually up to 40µm in diameter and about 1,000–10,000 times the volume of prokaryotes
DNA is a single circular molecule found free in the cytoplasm, DNA is 'naked'	DNA not circular and found in the nucleus which is surrounded by a membrane, DNA is associated with proteins (histones) to form chromosomes
Small ribosomes about 18nm in diameter (sometimes called '70S' ribosomes)	Ribosomes about 22nm in diameter (called '80S' ribosomes)
No endoplasmic reticulum present	Endoplasmic reticulum present plus Golgi body, chloroplasts and mitochondria, depending on species
Cell wall present	Cell walls present in plant cells only
Examples – bacteria and blue-green algae	Examples – animals, plants and protoctists

Use this table to draw and label a typical pro + eukaryote.

Animal and plant cells

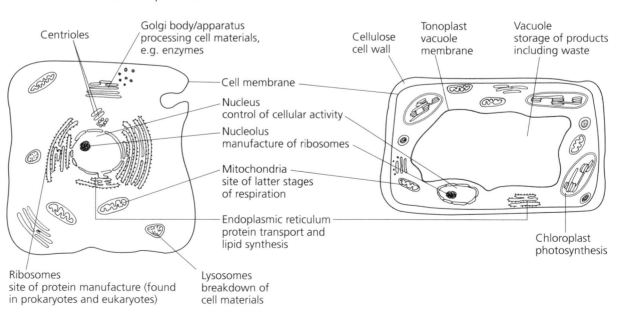

Centrioles

Golgi body/apparatus processing cell materials, e.g. enzymes

Cellulose cell wall

Tonoplast vacuole membrane

Vacuole storage of products including waste

Cell membrane

Nucleus control of cellular activity

Nucleolus manufacture of ribosomes

Mitochondria site of latter stages of respiration

Endoplasmic reticulum protein transport and lipid synthesis

Chloroplast photosynthesis

Ribosomes site of protein manufacture (found in prokaryotes and eukaryotes)

Lysosomes breakdown of cell materials

Movement in and out of cells

The cell surface membrane controls exchange between the cell and its environment. In 1972 Singer and Nicolson suggested a model for the structure of a cell membrane which they described as the **fluid-mosaic model**.

Phospholipid part is fluid with its molecules constantly moving about.

Uneven distribution of protein makes it look like a mosaic.

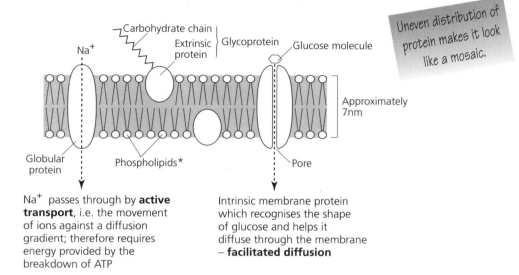

Na$^+$ passes through by **active transport**, i.e. the movement of ions against a diffusion gradient; therefore requires energy provided by the breakdown of ATP

Intrinsic membrane protein which recognises the shape of glucose and helps it diffuse through the membrane – **facilitated diffusion**

How does this structure compare with the general structure of a lipid?

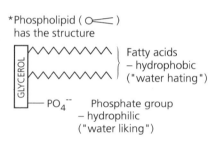

*Phospholipid (⊶)
has the structure

Fatty acids
– hydrophobic
("water hating")

PO$_4$$^{--}$ Phosphate group
– hydrophilic
("water liking")

Remember scale conversions (see glossary). It is better to work in whole numbers – start to use negatives and you'll make mistakes. Use a calculator and check your answer makes sense. If you divide by the magnification you'll get the real size.

The total thickness of the membrane averages 7nm.

The phospholipid bilayer around cells makes a very effective barrier, particularly against the movement of water soluble molecules and ions. The aqueous contents of the cell are therefore prevented from escaping. However, some exchange between the cell and its environment is essential.

Diffusion is the movement of molecules or ions from a region of **their higher concentration** to one of lower concentration, e.g. oxygen and carbon dioxide cross membranes by diffusion.

This always gives students problems in exams – try to write in terms of less negative and more negative rather than higher and lower, and you are less likely to be confused.

The inflow of water is resisted by the cell membrane (in plants also by the cell wall). This opposing force is called the pressure potential and it has a positive value. Water moves from regions of higher **water potential** (ψ) or more molecules of water to regions of lower water potential or fewer molecules of water. When this takes place through a **selectively permeable membrane**, this is called **osmosis**. Pure water has a water potential of 0 and the addition of a solute reduces the value, giving solutions negative values for ψ. Concentrated solutions have large negative water potentials; dilute solutions have small negative water potentials.

Equation

Water potential of cell	=	Water potential of cell sap	+	Pressure potential
ψ, usually -ve		ψS, always -ve		ψP, usually +ve

Some ions and molecules move across membranes by active transport, against a concentration gradient. This needs a carrier protein and ATP to provide energy, e.g. as in the sodium-potassium pump (see later).

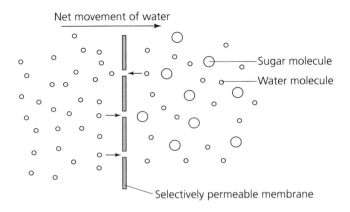

Net movement of water

Sugar molecule

Water molecule

Selectively permeable membrane

Osmosis in animal cells

Osmosis in plant cells

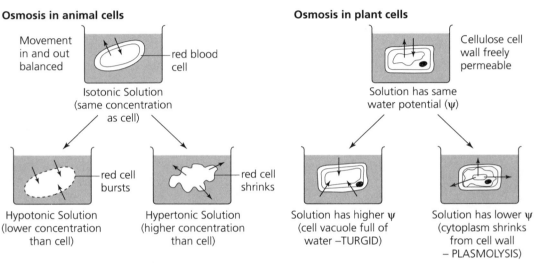

Movement in and out balanced — red blood cell

Isotonic Solution (same concentration as cell)

red cell bursts

Hypotonic Solution (lower concentration than cell)

red cell shrinks

Hypertonic Solution (higher concentration than cell)

Cellulose cell wall freely permeable

Solution has same water potential (ψ)

Solution has higher ψ (cell vacuole full of water –TURGID)

Solution has lower ψ (cytoplasm shrinks from cell wall – PLASMOLYSIS)

Try to think of osmosis in terms of water potential rather than concentration, i.e. as movement of water from high water potential to low water potential.

Autotrophic nutrition

All organisms need a source of energy. Respiration and photosynthesis are the two main methods used by organisms to make this energy available.

Both processes result, amongst other things, in the production of ATP.

Examiners still complain that candidates fail to appreciate that plants respire as well as photosynthesise.

> Energy released by the oxidation of glucose during respiration, or energy absorbed from sunlight during photosynthesis, can be used to make ATP. The energy helps to link a third phosphate group to a molecule of ADP and this is the energy which is made available to the cell when the bond is broken, i.e. when ATP is converted to ADP.

Autotrophic organisms (or 'self-feeding') make organic compounds from carbon dioxide. Most autotrophic organisms do this by photosynthesis. In the process of photosynthesis energy from sunlight is trapped by chlorophyll and used in the manufacture of carbohydrates from carbon dioxide and water. This process can be summarised:

Always remember this equation. It will help you identify the fate of the reactants and where the products come from.

$$CO_2 + 2H_2O \xrightarrow[\text{chlorophyll}]{\text{sunlight}} CH_2O + H_2O + O_2$$

carbon dioxide   water   carbohydrate   water   oxygen

Photosynthesis can be divided into two phases:

The **light-dependent stage** in which chlorophyll absorbs light energy which is then converted to chemical energy. Water molecules are split, oxygen is given off and the hydrogen is used to reduce carbon dioxide in the second stage. This stage needs light but is not affected by temperature.

The **light-independent stage** in which carbon dioxide is reduced and then converted into sugars by a series of reactions. This stage, because it involves chemical reactions, is affected by temperature. An increase in temperature of 10°C will approximately double the rate of reaction.

The equation can therefore be shown as follows:

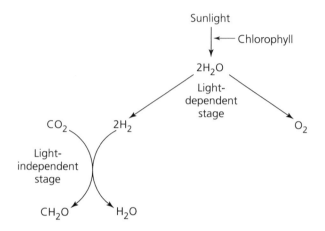

Chloroplasts are the cell organelles of green plants which carry out photosynthesis:

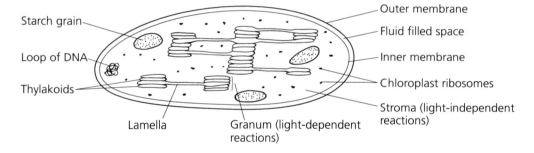

The biochemical evidence indicates that the thylakoid membranes contain all the components needed for absorbing light and carrying out the light-dependent reactions of photosynthesis, while the light-independent reactions take place in the stroma. Glucose made in these reactions is stored as starch grains, found in the chloroplasts.

Details of photosynthesis

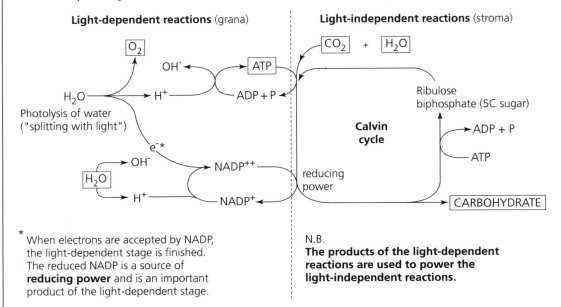

Light-dependent reactions (grana) | **Light-independent reactions** (stroma)

Photolysis of water ("splitting with light")

Calvin cycle

Ribulose biphosphate (5C sugar)

reducing power

CARBOHYDRATE

*
When electrons are accepted by NADP, the light-dependent stage is finished. The reduced NADP is a source of **reducing power** and is an important product of the light-dependent stage.

N.B.
The products of the light-dependent reactions are used to power the light-independent reactions.

Photosynthesis and temperature

The rate of photosynthesis initially increases with temperature. If a high temperature is maintained the rate declines, suggesting destruction of part of the system.

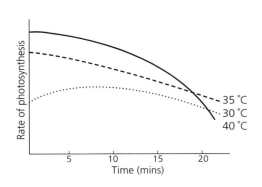

An exam question may ask you to plot any of these graphs from data and analyse them. Plot graphs carefully – many marks are lost by carelessness, such as using inappropriate scales.

The release of energy

Respiration is the process by which complex organic molecules are broken down to release energy:

$$C_6H_{12}O_6 + 6O_2 \longrightarrow 6CO_2 + 6H_2O + 2880 \text{ kJ/mol}$$

glucose | oxygen | carbon dioxide | water | energy

The number of carbon atoms is important. Always remember the starting point is a **6 carbon** molecule.

Respiration can be an **aerobic** process requiring the presence of oxygen to fully oxidise food materials. It will proceed in the absence of oxygen, i.e. **anaerobically**, but this produces less energy.

Aerobic respiration can be divided into 3 stages:

- glycolysis
- Krebs cycle/tricarboxylic acid cycle (TCA)
- electron transfer chain (ETC).

Krebs cycle and the electron transfer chain take place in the mitochondria of cells:

Mitochondrion

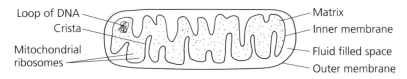

Loop of DNA
Crista
Mitochondrial ribosomes
Matrix
Inner membrane
Fluid filled space
Outer membrane

Glycolysis (glucose splitting)

Check your syllabus to see if you need to know the number of ATP molecules produced.

- The first stage of respiration.

- Common to both aerobic and anaerobic respiration.

- Takes place in the cell **cytoplasm**.

- 2 ATP molecules are consumed in initial **phosphorylation** (adding phosphate) reactions.

- 4 ATP molecules are produced.

- Net gain of 2 ATP molecules per glucose molecule.

- 4 atoms of hydrogen transported to electron transfer chain as 2 molecules of $NADH_2$.

- 2 molecules of pyruvic acid produced for every glucose molecule.

Summary

Keep counting the carbon atoms. Remember one molecule of glucose produces 2 molecules of pyruvate.

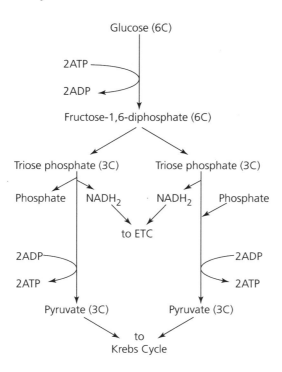

Glucose is phosphorylated twice i.e. 2 phosphate groups are added, each coming from a molecule of ATP.

Fructose –1,6-diphosphate is split into 2 3-carbon phosphates.

Both triose phosphates are oxidised (H_2O removed) and phosphorylated (phosphate added).

Series of reactions removing phosphate groups to produce 2 molecules of ATP.

Final product is **2 molecules of pyruvate and 2 molecules of ATP per glucose molecule**.

Krebs Cycle

- The reactions of Krebs Cycle occur in the matrix of the mitochondria.

- Before pyruvate can enter the cycle it must react with coenzyme A to form **acetyl coenzyme A** – this is the link stage between glycolysis and Krebs Cycle.

- The **decarboxylation** reactions remove CO_2 which will eventually be exhaled.

- 1 molecule of ATP is generated directly from each turn of the cycle.

- The cycle of reactions is completed twice for each molecule of glucose.

- The link stage and Krebs Cycle together produce:
  - 2 molecules of ATP
  - 10 pairs of hydrogen atoms to go into the electron transfer chain
  - 6 molecules of CO_2,
  ᵧ for every glucose molecule oxidised.

- Krebs Cycle is also concerned with the oxidation of fats and proteins – they enter the cycle via acetyl CoA.

Remember – twice around the cycle for every glucose molecule.

Summary

If you remember the simplified equation, you'll be able to show where the products come from.

Check your syllabus to find out how much detail is needed – keep counting the carbons!

Pyruvate (3C)

CO_2 ← | → $2H^+$

Acetyl CoA (2C)

4C acid now ready to combine with another acetyl CoA and cycle repeats.

4C Acid → Citrate (4C)

$2H^+$

$2H^+$

Pyruvate undergoes oxidative decarboxylation and combination with coenzyme A to form acetyl coenzyme A.

Acetyl CoA combines with 4C acid to form citrate and CoA returns to 'pick up' another pyruvate molecule.

CO_2

$2H^+$

$2H^+$

5C acid is converted to 4C acid in 3 steps, releasing 2 pairs of H atoms.

5C Acid

CO_2

ATP

Another oxidative decarboxylation forms a 5C acid and enough energy is released to generate a molecule of ATP.

Electron transfer chain

- The electron transfer chain is a chain of carrier molecules.

If you cannot remember any of the chemistry remember this, it is very important.

- **The carriers are alternately reduced and oxidised** as the electrons are transferred from one carrier to the next to **release energy** as electrons move to lower energy levels.

- The components of the chain are in the lipid bilayer of the inner mitochondrial membrane.

- The movement of protons across the membrane back into the matrix drives ATP synthesis – every 2 protons generate 1 molecule of ATP.

- Hydrogen atoms are brought to the chain by $NADH_2$ or $FADH_2$ as protons and electrons.

Remember
Oxidation
Is
Loss (of electrons)
Reduction
Is
Gain

- Oxygen serves as the final hydrogen acceptor, combining to form water – see what this is for the simplified equation.

- Without oxygen all the respiratory stages would soon become stuck in their reduced state and production would stop.

Try reading the above and summarising it in your own diagram.

N.B.

Hydrogen is transported to the electron transfer chain by the hydrogen carriers NAD and FAD or Nicotinamide Adenine Dinucleotide and Flavin Adenine Dinucleotide. These reduced molecules are the source of **reducing power** which can be used to make ATP in the electron transfer chain.

Anaerobic respiration

- Anaerobic respiration occurs when there is insufficient oxygen for aerobic respiration to take place.

- Of the 3 main respiratory stages, only glycolysis operates.

- In animals the pyruvate is reduced to **lactic acid**.

Check your syllabus to find out if you need to know the chemistry of this.

- When pyruvic acid instead of oxygen accepts the hydrogen atoms from $NADH_2$, the body is said to be building up an **oxygen debt**.

- in plants, the pyruvate is reduced to **ethanol**, and **carbon dioxide** is liberated – this is called **alcoholic fermentation**.

- Compared with aerobic respiration, anaerobic respiration is an inefficient process producing only 2 ATP molecules per molecule of glucose, i.e. about 1/20 of total yield.

Suggested further reading

Alberts, B. (ed.) (1989) *Molecular Biology of the Cell*, Garland Publishing Co. Inc, New York (0-8240-3696-4)

Aldridge, S. (1993) *Biochemistry: A Textbook for A level Biology*, Cambridge University Press (0-521-43781-4)

Carr, M. & Cordell, B. (1992) *Biochemistry (Biology – Advanced Studies)*, Nelson (0-17-44819-9)

James, D. & Matthews, G. (1993) *Understanding the biochemistry of respiration*, Cambridge University Press (0-521-39993-9)

Hall, D.O. & Rao, K.K. (1994) *Photosynthesis* (5th edition), Cambridge University Press (0-521-43622-2)

The Biochemical Society – *Biochemistry Across the Curriculum Series*
    Number 1 – *Essential Chemistry for Biochemistry*
    Number 3 – *Enzymes and their role in Biotechnology*
    Number 4 – *Metabolism*
    Number 6 – *Photosynthesis*

Chapter two

Continuity of life

Glossary

Allele – one of a number of alternative forms of a gene, each possessing a unique nucleotide sequence, only one of which can appear at a locus (the position of the gene in the DNA molecule)

Chromosome – a length of DNA consisting of a number of **genes**

Continuous variation – a character is said to vary continously if individuals show a range of phenotypes with a smooth graduation from one extreme to the other, rather than falling into a number of discrete categories, e.g. height in humans

Dihybrid – the inheritance of **two** characteristics

Diploid – the total number of chromosomes in a body cell nucleus, e.g. in humans the diploid number is 46

Discontinuous variation – characters within a population vary discontinuously if they exhibit a limited form of variation producing individuals with clear-cut differences but no intermediates between them, e.g. blood groups in humans

DNA – deoxyribonucleic acid; double helix found in the nucleus which carries the genetic code

Dominant – the condition in which the effect of an allele is expressed in the phenotype, even in the presence of an alternative (recessive) allele

Gene – a specific length of DNA made up of a sequence of nucleotides to which a specific function can be assigned

Genotype – the genetic constitution of an organism with respect to the alleles under consideration

Haploid – the number of chromosomes found in the gametes of an organism, e.g. in humans the haploid number is 23

Heterozygous – situation where the alleles are different, e.g. Aa

Homozygous – situation where the alleles are both the same, e.g. AA or aa

Meiosis/sexual reproduction – form of cell division which results in the production of genetically varied sex cells or **gametes**

Mitosis/asexual reproduction/growth – form of cell division in which the daughter cells produced are genetically identical to the parent cell

Monohybrid – the inheritance of **one** characteristic

Mutation – change in the genetic information of a cell, which can be inherited by subsequent generations

Phenotype – the characteristics of an individual, usually resulting from the interaction between the genotype and the environment

Recessive – the condition in which the effect of an allele is expressed in the phenotype of a diploid organism only in the presence of another identical allele

RNA – ribonucleic acid; found in a number of forms, messenger, transfer and ribosomal; differs from DNA in being a single strand

Transcription – the copying of the coded message from DNA to messenger RNA (mRNA)

Translation – the decoding or translation of the message on the mRNA into a polypeptide chain

Zygote – cell resulting from the fusion (**fertilisation**) of gametes

Cell division

Cells divide on a regular basis to bring about growth and asexual reproduction. Asexual reproduction and growth are the result of a type of cell division called **mitosis**.

Mitosis produces two identical daughter cells with genetic material identical to the parent. Each species has a characteristic number of **chromosomes**, e.g. humans have 46 comprising 23 **homologous pairs** (known as the **diploid number**).

The **cell cycle** is the period of time incorporating division by mitosis and the non-dividing **interphase** between divisions, which can last from 24 hours to years depending upon species.

Mitosis

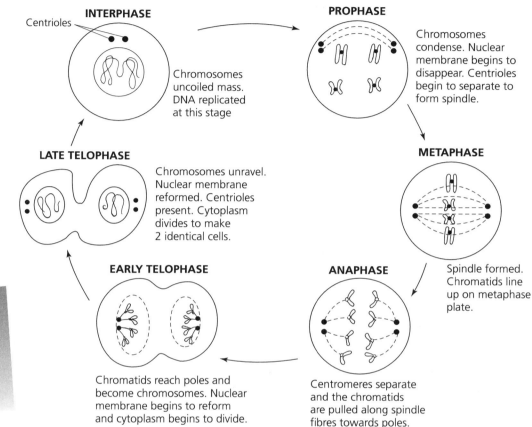

If you are asked to draw and write about mitosis in an exam use 2 chromosome pairs in your example – it's less confusing.

Gametes are specialised cells which have half the number of chromosomes (the **haploid number**), e.g human gametes (eggs and sperm) have 23 chromosomes, or one from each homologous pair. In sexual reproduction two haploid nuclei fuse to form a **zygote** which then becomes diploid. Gametes are formed in the gonads as a result of a **reduction division** or **meiosis**.

Structure of chromosomes

- Chromosomes are made up of DNA and carry the blueprint for the proteins which determine the make up of the cell and the entire organism.

- Each chromosome contains one molecule of deoxyribonucleic acid (DNA).

Examiners still complain that candidates confuse chromosomes and genes – the short word represents the shorter structure!

The nature of the gene

- The basic structure and working parts of most cells are made up of carbohydrates, proteins and lipids.

- The control of their function is the job of another group of biological molecules – the **nucleic acids** (DNA and RNA)

Meiosis

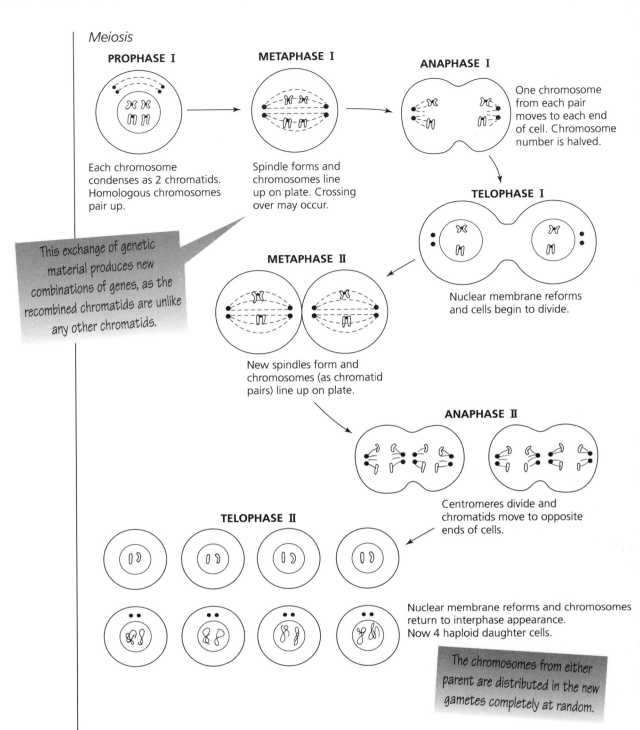

PROPHASE I

Each chromosome condenses as 2 chromatids. Homologous chromosomes pair up.

METAPHASE I

Spindle forms and chromosomes line up on plate. Crossing over may occur.

ANAPHASE I

One chromosome from each pair moves to each end of cell. Chromosome number is halved.

This exchange of genetic material produces new combinations of genes, as the recombined chromatids are unlike any other chromatids.

TELOPHASE I

Nuclear membrane reforms and cells begin to divide.

METAPHASE II

New spindles form and chromosomes (as chromatid pairs) line up on plate.

ANAPHASE II

Centromeres divide and chromatids move to opposite ends of cells.

TELOPHASE II

Nuclear membrane reforms and chromosomes return to interphase appearance. Now 4 haploid daughter cells.

The chromosomes from either parent are distributed in the new gametes completely at random.

DNA

The structure of DNA

DNA is a helical molecule, like a ladder that has been twisted.

The sides of the ladder are made of alternate sugar (deoxyribose) and phosphate molecules

The rungs of the ladder are pairs of nitrogenous bases

The four DNA bases are
– 2 purines, adenine (A) and guanine (G)
– 2 pyrimidines, cytosine (C) and thymine (T)

A always pairs with T.
C always pairs with G.

Each sugar molecule with its phosphate and attached base is called a **nucleotide**.

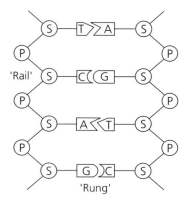

The DNA 'ladder'

'Rail'

'Rung'

A nucleotide

A nucleotide consists
of a base, a sugar and
a phosphate

Base pairing

Bases pair purine with pyrimidine

[T > A]——Thymine and adenine

[C(G]——Cystosine and guanine

DNA replication

- To be sure that the same genetic material is passed on to all new cells, DNA needs to produce an exact copy of itself, i.e. **replicate**.

- Watson and Crick suggested that this would occur if the two strands of DNA separated and each one acted as a template for making a complementary strand.

DNA replication (the 'zip-fastener' idea)

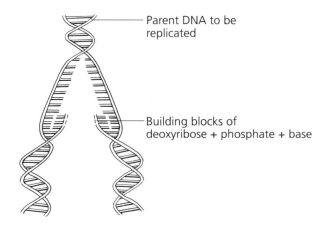

Parent DNA to be replicated

Building blocks of deoxyribose + phosphate + base

This is known as the **semi-conservative theory of replication** and was demonstrated by Michelson and Stahl in 1958.

If the 'zip-fastener' hypothesis was correct Michelson and Stahl suggested that neither of the products of DNA replication should be completely new – in both daughter DNA molecules one of the two strands would be new, while the other would be one of the parental strands. Alternatively, if the 'conservative' theory is correct, one of the two daughter DNAs would be completely new, while the other would be the original parent which had somehow stimulated the synthesis of a second double helix identical to itself.

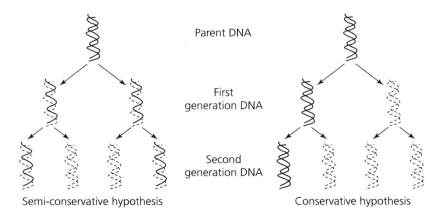

Parent DNA

First generation DNA

Second generation DNA

Semi-conservative hypothesis

Conservative hypothesis

Michelson and Stahl designed an experiment to determine which hypothesis was correct.

Cells of *E.coli* were grown for many generations on a medium in which normal nitrogen (^{14}N) was replaced with the heavy isotope, ^{15}N. Once enough time had passed for most of the nitrogen atoms in the DNA molecules of *E.coli* to be of the heavy type, the bacteria were introduced into a new medium containing normal nitrogen. Samples of bacteria were then withdrawn at intervals equal to the generation time, and the relative amounts of the two types of nitrogen estimated by a technique which relied on the fact that molecules containing ^{15}N are very slightly heavier than those containing normal nitrogen. Ultracentrifugation was used to separate the DNA molecules according to the ratio of ^{14}N to ^{15}N they contained.

The results (shown below) support the semi-conservative (zip-fastener) hypothesis. In the first generation after being switched back to normal ^{14}N, the DNA was found to have a density midway between what it would have if it contained only ^{14}N or only ^{15}N; in other words it contained equal amounts of each.

In the second generation, two sorts of DNA were detected; one sort contained only ^{14}N; the other sort was the same as that obtained in the first generation, i.e. it contained equal amounts of ^{14}N and ^{15}N. These results fit perfectly with what would be expected if DNA replicated by the semi-conservative hypothesis.

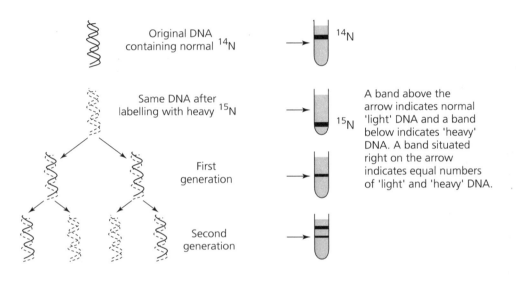

Original DNA containing normal ^{14}N — ^{14}N

Same DNA after labelling with heavy ^{15}N — ^{15}N

A band above the arrow indicates normal 'light' DNA and a band below indicates 'heavy' DNA. A band situated right on the arrow indicates equal numbers of 'light' and 'heavy' DNA.

First generation

Second generation

A popular exam question is to ask for an interpretation of the results – use diagrams, it will make your explanation clearer and save time and writing.

DNA and protein synthesis

The following describes how DNA codes for amino acids.

- DNA instructs cells to make specific proteins.

- Information in the DNA determines the sequence of amino acids in the protein.

- This information is the sequence of bases in DNA and is known as the **genetic code**.

- Proteins are made of up to 20 different amino acids.

- DNA contains only four bases.

- For four bases to form an 'alphabet' for the 20 different amino acids the bases must code in groups of 3, called **triplet codons**.

- Protein synthesis requires a supply of amino acids, energy (from ATP), DNA and another nucleic acid called **ribonucleic acid** or **RNA**.

- RNA is found in the nucleus but most is found in the cytoplasm, especially in the ribosomes. Its sugar is ribose and its bases are the same as DNA, except that uracil (U) replaces thymine. RNA can be found as **messenger RNA** (**m RNA**), **transfer RNA** (**t RNA**) and **ribosomal RNA** (**r RNA**) and it is single-stranded.

Examiners continue to point out a common mistake – DNA is not a protein. Remember DNA is built up from nucleotides and proteins are built up from amino acids.

Summary of protein synthesis

DNA **transcription** to mRNA ① ⟶ mRNA ⟶ Ribosomes ② ⟶ **Translation** ⟶ Polypeptides and proteins

Nucleus

Amino acid 'pool'

Transfer RNA ③

Cytoplasm

① DNA molecule carries the code for a protein but is unable to leave the nucleus and so must form a template which is transcribed on to messenger RNA.

② mRNA is a single-stranded nucleic acid which can leave the nucleus carrying the information for protein synthesis. It is made as it is needed.

③ Transfer RNA carries a 'matching code' for mRNA in the form of triplets of bases. Each triplet is a match for the amino acid it also carries. Each amino acid is carried to a ribosome where it is joined (polymerised) to others to form a polypeptide chain.

> It has to be in 2 parts because the DNA is 'trapped' in the nucleus and synthesis takes place in the cytoplasm.

> Keep in mind the different structures of DNA and RNA and this will help you answer questions about their differing functions.

Protein synthesis can be divided into 2 main parts – **transcription** and **translation**.

Summary of transcription

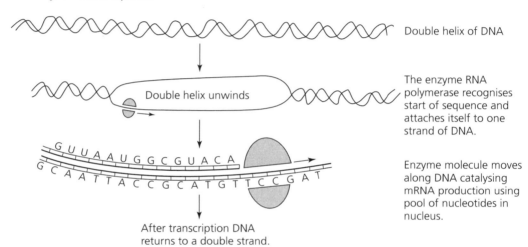

Double helix of DNA

Double helix unwinds

The enzyme RNA polymerase recognises start of sequence and attaches itself to one strand of DNA.

Enzyme molecule moves along DNA catalysing mRNA production using pool of nucleotides in nucleus.

G U U A A U G G C G U A C A
G C A A T T A C C G C A T G T T C C G A T

After transcription DNA returns to a double strand.

> Questions which ask about transcription are asking you to show you understand the mRNA sequence which corresponds to the DNA. Set out your answer carefully giving yourself plenty of space:
> A A C – DNA
> U U G – mRNA
> and you won't go wrong!

Summary of translation

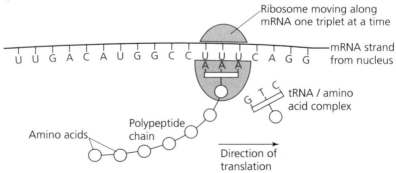

Ribosome moving along mRNA one triplet at a time

mRNA strand from nucleus

U U G A C A U G G C C U U U C A G G

tRNA / amino acid complex

Amino acids

Polypeptide chain

Direction of translation

Mutation

- Mutations are changes in the chemical structure of an individual gene or in the physical make-up of a chromosome.
- Mutations are a source of variation.
- Most mutations are not beneficial.
- There are a number of different types of mutation.
- Changes in an individual gene are called **point mutations**.
- Changes in the position of genes on a chromosome are called **chromosomal mutations**.

> You won't mix up the types of mutation if you consider chromosome mutations as 'physical' changes and point or gene mutations as chemical changes.

Summary of mutations

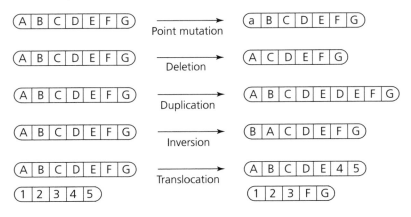

Genetic engineering

Genetic engineering involves moving from one organism to another of the same or different species, resulting in altered organisms.

The process involves cutting out the gene to be altered using highly specialised enzymes called **restriction endonucleases**, and adding the required gene in its place.

Genetic engineering is used in the production of human insulin by bacteria. In the future, genetic disorders may be treated by genetic engineering (See Biotechnology section for more detail).

Mendel and the laws of heredity

Gregor Mendel was a Czechoslovakian monk who carried out the first quantitative experiments on heredity using garden peas.

The first set of experiments involved the inheritance of a single pair of contrasting characteristics – **monohybrid inheritance**.

Mendel's first experiment showing monohybrid inheritance

He crossed pea plants producing wrinkled seeds with ones which produced round seeds.

To represent characteristics, use letters which have a clear difference between the capital and small letter. If this is not possible draw a line through the small letter e.g. S and $.

Write the gamete genotypes in a circle and remember each gamete can only have one allele from each pair.

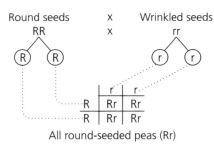

Round seeds	x	Wrinkled seeds	Parent phenotypes (P$_2$)
RR	x	rr	Parental genotypes (2n)
			Meiosis
Ⓡ Ⓡ		ⓡ ⓡ	Gametes (n)

	r	r
R	Rr	Rr
R	Rr	Rr

Random fertilisation

All round-seeded peas (Rr)　　First generation F$_1$

Self-pollinated Rr plants

Take care setting out genetics problems – always use side headings or annotations to explain what you have done, e.g. parental phenotypes, parental genotypes, etc.

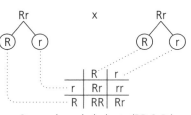

| Rr | x | Rr | Second parental generation (P$_2$) |
| Ⓡ ⓡ | | Ⓡ ⓡ | |

	R	r
r	Rr	rr
R	RR	Rr

Second generation F$_2$

3 round-seeded plants (RR & Rr)
1 wrinkled-seeded plant (rr)

From this it was concluded that:

- inheritance is 'particulate' (we now know these 'particles' as genes)
- genes occur in pairs which may be **dominant** or **recessive** with respect to each other
- one gene can be carried by a single gamete

Rarely asked for directly in an exam but an understanding is often tested.

Mendel's First Law – the **Law of Segregation** states:

Think how this corresponds to the movement of chromosomes.

'The characters of an organism are controlled by pairs of alleles which separate in equal numbers into different gametes as a result of meiosis.'

An organism's characteristics or **phenotype** are determined, in part, by its genetic constitution or **genotype**.

What emerged from Mendel's experiments was that a particular phenotype may be produced by more than one genotype, e.g. a tall pea plant may be TT or **homozygous**, or Tt or **heterozygous**. This may be tested by carrying out a **test cross** or **back cross**.

Test cross

Usually forms the final part of a question. **Remember** you can only 'test' against something you know and the only thing you know for certain is the homozygous recessive.

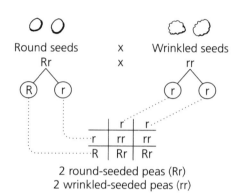

Sometimes Mendel's Law does not seem to apply.

Incomplete dominance

This is often associated with sex-linked characteristics. A favourite in exam questions is coat colour in cats.

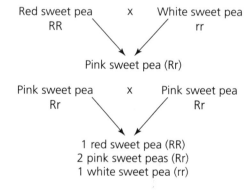

Further experiments involved Mendel in studying the inheritance of 2 pairs of characteristics – **dihybrid inheritance**.

Mendel's experiment to show dihybrid inheritance

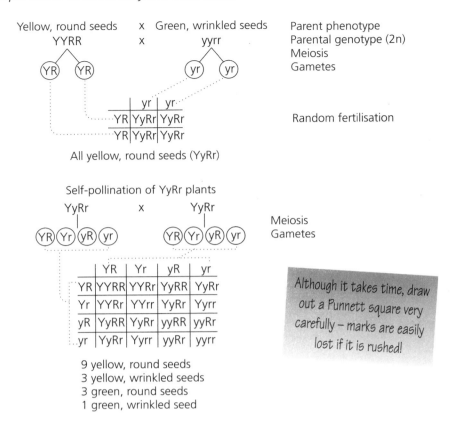

From this Mendel concluded that genes segregate independently when gametes are formed.

Mendel's Second Law – the **Law of Independent Assortment** states:

'The alleles of genes at different loci segregate independently of one another during the formation of gametes.'

As with monohybrid inheritance, a test cross will establish an organism's genotype.

Genes carried on the X chromosome are said to be **sex-linked**, e.g. haemophilia and colour blindness in humans. Genes such as the one causing colour blindness are recessive resulting in a much greater chance of a male inheriting the condition than a female:

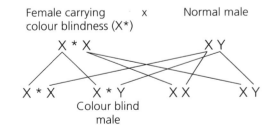

- Linked genes can be separated as a result of **chiasmata** forming during meiosis and **crossing over** taking place.

- Certain characteristics are controlled by **multiple alleles**, such as the ABO blood group system in humans.

- **Lethal genes** usually result from a particular combination of genes.

Make sure you understand the difference between genes and alleles. There is a gene for eye colour with alleles for blue, brown, etc.

This can also be drawn as a Punnett square. Use whichever you feel most comfortable with – examiners accept both.

Although it takes time, draw out a Punnett square very carefully – marks are easily lost if it is rushed!

Make sure you mark the carrier carefully so you don't become confused and make a silly mistake.

Summary of crosses and their ratios

Type of cross		Example	Expected ratio
Phenotype	Genotype		
Tall x tall plants	Tt x Tt	Peas – heterozygous tall plants self-pollinated	3:1, tall : dwarf
Tall x dwarf plants	Tt x tt	Peas – heterozygous tall x homozygous dwarf **test cross**	1:1, tall : dwarf
Pink x pink flowers	Rr x Rr	Heterozygous pink flowers self-pollinated **incomplete dominance**	1:2:1, red : pink : white
Round, yellow seeds x wrinkled, green seeds	RrYy x RrYy	Peas – heterozygous round, yellow plants self-pollinated	9:3:3:1, yellow, round : yellow wrinkled : green round : green wrinkled
Round, yellow seeds x wrinkled, green seeds	RrYr x rryy	Heterozygous round, yellow plants x homozygous wrinkled, green plants **test cross**	1:1:1:1, yellow, round : yellow, wrinkled : green round : green wrinkled

These are just a few examples. Try to think of some yourself. It will help you learn the ratios and give you a wider range to choose from when answering an exam question.

Evolution

There are approximately 6,000 million people on the earth and every one is different (except identical twins).

Some differences between individuals of the same species are brought about by variations in the environment, e.g. coat colour in mammals – the arctic hare changes from brown to white in the winter. The environment induces changes in the activity of the genes controlling coat colour, but the actual genes present are the same in both brown and white individuals.

Genetically different individuals maintain their different characters whatever the environmental conditions, e.g. melanin in the skin which gives protection against the sun. The degree of pigmentation varies according to exposure to the sun. However, even in the same environment, there is always a pigmentation difference between Negroid and Caucasian individuals. This is because the combinations of genes controlling the distribution of melanin are different.

Discontinuous variation is the kind where there are distinct categories, e.g. blood groups or sex. An organism is one or the other due to the genes present, environment has little effect.

Continuous variation is the kind with many intermediate categories and is influenced by environment as well as the genes present, e.g. height.

Breeding programmes are a common part of agricultural research, horticulture, etc. This is called **artificial selection**.

Nature's 'breeding programme' is called **natural selection**.

Make sure you know about other theories, e.g. Lamarck.

Evolution describes a very long series of gradual changes which have taken place, resulting in the vast range of present-day species. The idea of natural selection as a mechanism for evolution was introduced by Charles Darwin in 1859.

Evidence for evolution comes from studies into the geographic distribution of organisms, comparative anatomy, taxonomy, embryology, cell biology and palaeontology.

Learn an example to illustrate each piece of evidence.

Recent developments of the theory take into account present-day knowledge of genetics:

Mendelian inheritance – characteristics handed on to the next generation.

↓

Sexual reproduction (meiosis) results in considerably more variation (both continuous and discontinuous) than is accounted for by mutation.

↓

Genes occasionally come together and result in an organism more suited to the environment and which will, therefore, have a better chance of surviving and reproducing – natural selection or the 'survival of the fittest'.

↓

Over time these 'selected' changes may result in a completely different species.

Be able to explain 'survival of the fittest' with clear reference to what the organism is 'fitting into'.

Population genetics

Effects of predation, competition etc. are important – see ecology.

A population is a group of individuals of the same species occupying a particular habitat.

The gene pool is the sum total of the genes in a population at a given time.

Check with your syllabus that you need to know this.

> If p represents the frequency of a dominant allele and q the frequency of a recessive allele, then $p + q = 1$. p and q are impossible to measure because heterozygous and homozygous dominants have the same phenotype.
>
> The Hardy-Weinberg equation can be solved to give the genotypes by observing the number of individuals with the recessive phenotype (homozygous recessives):
>
> $p + 2pq^2 + q = 1$
>
> Hardy-Weinberg equilibrium only occurs where the gene frequencies are stable, i.e. the population is not evolving – there is random mating, no immigration or emigration, no mutations occur and the population is large.

Types of selection

- **Natural selection** results in an increase in the proportion of advantageous alleles in a population.

- **Stabilising selection** maintains advantageous characteristics already present in the population.

- **Directional selection** results in a change to a new phenotype which is better suited to a changing environment.

- **Diversifying selection** results in different sub-populations with their own phenotypes better suited to different habitats.

- **Balancing selection** maintains variety, where a disadvantageous allele is kept in a population, e.g. by being advantageous in the heterozygous state (sickle cell trait).

Suggested further reading

Burnet, L. (1993) *Essential Genetics*, Cambridge University Press (0-521-31380-5)

Carter, M. (1992) *Genetics and Evolution*, Hodder and Stoughton (0-340-53266-1)

Clarke, C.A. (1987) *Human Genetics and Medicine* (3rd edn), Edward Arnold (0-7131-2944-1)

Clark, B.C.F. (1992) *The Genetic Code and Protein Synthesis* (2nd edn) Studies in Biology Number 83, Cambridge University Press (0-521-42763-0)

Dawkins, R. (1989) *The Selfish Gene*, Oxford University Press (0-19-286092-5)

Dawkins, R. (1991) *The Blind Watchmaker*, Penguin (0-14-14481-1)

Gonick, L. & Wheelis, M. (1983) *The Cartoon Guide to Genetics*, Barnes and Noble Books (0-06-460416-0)

Nicholl, D.S.T. (1994) *An Introduction to Genetic Engineering*, Cambridge University Press (0-521-43634-6)

Tomkins, S. (1992) *Heredity and Human Diversity*, Cambridge University Press (0-521-31229-9)

The Biochemical Society – *Biochemistry across the Curriculum* No. 2 – *The Structure and Function of Nucleic Acids*

Chapter three

Physiology of animals and plants

Glossary

Carnivores – feed on flesh

Cohesion-adhesion – characteristics of water molecules which help explain water movement in plants

Diastole – relaxation

Diffusion – the transfer of substances from places of high concentration to places of low concentration (net movement occurs because atoms and molecules are in a state of continual random motion, and continues until a uniform concentration is established in the space available

Halophytes – seashore dwelling plants

Herbivores – feed on plant material only

Heterotrophic – literally **hetero** – 'other', **trophic** – 'feeding'

Holozoic – feeds on solid organic material

Hydrophytes – water dwelling plants

kPa – kiloPascal – unit of pressure

Lymphocyte – white blood cell which produces antibodies

Mass flow – possible mechanism of sucrose transport in plants

Mesophytes – land dwelling plants

Myogenic – spontaneous rhythmic beating of the heart

Parasitic – feeds on another living organism known as the host

Phagocyte – cell that ingests pathogens, literally **phagy** – 'eating', **cyte** – 'cell'

Saprophytic – feeds on the solid organic materials from dead organisms

Selective reabsorption – the selective uptake of solute molecules and water in amounts useful to the body

Systole – contraction (hint – s̲ystole / s̲queeze)

Translocation – movement of organic solutes in a plant

Transpiration – loss of water from the leaves and aerial parts of a plant

Tubular secretion – occurs by active transport and removes solutes from body fluids to the filtrate or directly to the environment. It operates in the opposite direction to reabsorption.

Ultra-filtration – the process by which solvent and solute separate from solution according to their ability to pass through pores in a filter. In the body the force required to produce filtration is blood pressure

Vasoconstriction – narrowing of blood vessels to restrict blood flow

Vasodilation – expansion of blood vessels to increase blood flow

Xerophytes – desert dwelling plants

Gaseous exchange in animals and plants

Make sure you know the difference between breathing, gaseous exchange and respiration. These are often confused in answering questions.

Many living organisms (both plants and animals) are single cells where simple diffusion is sufficient to supply oxygen and remove carbon dioxide. The surface area of a cell compared with its volume (surface area:volume ratio) is relatively large for single cells and small organisms, so diffusion can be effective. However, for larger organisms the SA:Vol ratio changes, so that diffusion is not adequate and cells need to become specialised to facilitate gaseous exchange.

Animals

Respiratory systems have evolved to enable animals to obtain oxygen and remove carbon dioxide from their bodies – a process known as gaseous exchange.

Gaseous exchange takes place at respiratory surfaces which:
- have a large surface area
- are permeable
- are thin
- are moist
- have a rich blood supply.

Composition of breathed air (approx. %)

	nitrogen	oxygen	carbon dioxide
inspired air	79	20.96	0.04
expired air	79.6	16.4	4

The muscles used for breathing are 'voluntary', i.e. can be consciously operated, for example during singing or playing a wind instrument.

This action can be overridden by the respiratory centre in the brain.

Plants

The living cells of a flowering plant require oxygen for aerobic respiration. During daylight the rate of photosynthesis exceeds the rate of respiration and, therefore, the net flow of gases is towards intake of carbon dioxide and removal of oxygen. There are no specialised systems within plants for obtaining oxygen and removing carbon dioxide, although leaves are **adapted** for gaseous exchange.

The **stomata** connected to large air spaces in the spongy mesophyll layer of leaves, together with openings in woody stems called **lenticels**, ensure gaseous exchange takes place.

Gases move in and out through stomata

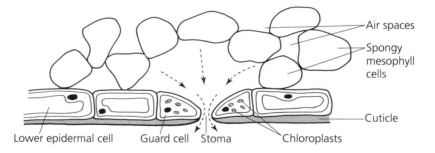

Air spaces

Spongy mesophyll cells

Cuticle

Lower epidermal cell   Guard cell   Stoma   Chloroplasts

Heterotrophic nutrition

'Heterotrophic' literally means 'feeding on others'.

There are three main types of nutrition.

- **Holozoic nutrition** – feeding on solid organic material from living or dead bodies of other organisms (either plants or animals). This is seen in animals and some carnivorous plants.

- **Parasitic nutrition** – feeding from the body of another living organism known as the host. This is seen in both animals and plants.

- **Saprophytic nutrition** – feeding on soluble organic materials from dead organisms, e.g. fungi.

Make sure you can draw and annotate a diagram of the lungs and respiratory system

What are the consequences of plants respiring **and** photosynthesising?

What role does diffusion play? How is this affected by weather conditions?

A heterotrophic organism has to be able to obtain, digest and absorb food.

Heterotrophs which rely on eating plants – **herbivores** – do not have to catch their food. Digestion can be complex because of the cellulose cell walls of plant cells. Heterotrophs which kill and eat other animals – **carnivores** – have an easier time in digestion but their food can be difficult to catch!

Digestion of food is the breakdown of complex molecules into simple soluble ones that can be absorbed.

Summary of digestion and absorption

Food – large, insoluble molecules

Enzymes added (also chemicals such as bile salts which emulsify fats)

Food broken down into soluble molecules

Digested food absorbed

Water absorbed

Faeces – undigested food

When revising, draw a large version of this simple plan and label enzymes, etc. on it.

Without looking back, define a catalyst and explain the lock and key hypothesis of enzyme action.

Chemical digestion in a mammal is catalysed by enzymes and takes place in the alimentary canal.

Make sure you can annotate a diagram of the digestive system

Summary of chemical digestion

Substrate	Product	Enzyme	Produced by . . .	Site of action
Polysaccharide e.g. starch	Disaccharides e.g. maltose	Salivary amylase	Salivary glands	Mouth
Maltose	Glucose	Maltose	Small intestine wall	Small intestine
Sucrose	Glucose + fructose	Sucrose	Small intestine wall	Small intestine
Lactose	Glucose + galactose	Lactose	Small intestine wall	Small intestine
Protein	Polypeptides	Pepsin	Pepsinogen –> pepsin in presence of acid from stomach wall	Stomach
		Trypsin	Pancreas	Small intestine
	Polypeptides	Endopeptidases	Pancreas	Duodenum
Peptides	Amino acids	Exopeptidases	Pancreas	Duodenum
Casein	Coagulated protein	Rennin	Young mammals only – stomach	Stomach
Fats	Fatty acids and glycerol	Lipase	Pancreas	Duodenum

The terminology is linked to what the enzymes do. For example, you need to know endo/exo peptidase + where these act.

Uptake and transport in plants

If you have difficulty remembering what each vessel transports, remember phloem (f) – food!

A transport system has evolved in higher plants to move substances around quickly.

The specific structures are **xylem** for transporting water and minerals up from the roots, and **phloem** for transporting the soluble products of photosynthesis from the leaves.

Arrangement of vascular bundles in a stem

Xylem is on the inside – it becomes the woody part of the plant. Phloem stays alive on the outside.

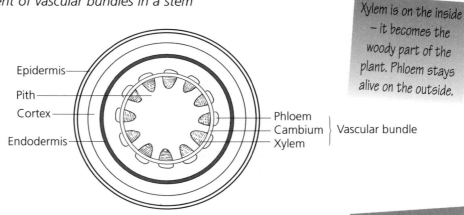

Epidermis
Pith
Cortex
Endodermis
Phloem
Cambium — Vascular bundle
Xylem

What experiment demonstrates root pressure?

Xylem vessels are dead tubes about 0.01–0.02mm in diameter. The movement of water in the xylem vessels depends on a number of factors, such as root pressure and a process called **transpiration**.

Transpiration is the loss of water vapour from the surface of a plant – 90% from stomata and 10% through the leaf cuticle. The amount of water lost by a plant due to transpiration can be very large, e.g. an oak tree can lose up to 600 dm^3 per day.

What external conditions produce the best diffusion gradient?

Water moves from the xylem into the mesophyll and evaporation takes place into the inter-cell spaces. Water vapour then moves from the cell spaces through the open stomata into the external air along a diffusion gradient.

A leaf has a layer of still air around it and the thickness of the layer depends on the surrounding wind and its speed. Transpiration is affected by light, temperature, air movement, air humidity and the availability of water in the soil. A **potometer** is an instrument used for measuring water uptake and the factors which affect it.

A potometer for measuring transpiration

What precautions must be taken when setting up a potometer?

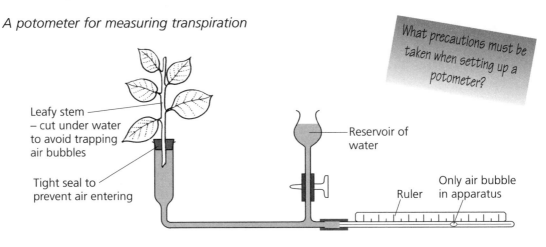

Leafy stem – cut under water to avoid trapping air bubbles

Tight seal to prevent air entering

Reservoir of water

Ruler

Only air bubble in apparatus

The **transpiration stream** is the continual movement of water up through the xylem, caused by the cohesion of water molecules (their tendency to 'stick together') and adhesion between water molecules and the walls of the narrow xylem vessels.

The combined cohesion-adhesion allows the column of water to be pulled upwards. More water is continually moved into the roots to replace that lost by transpiration.

Summary of transpiration stream

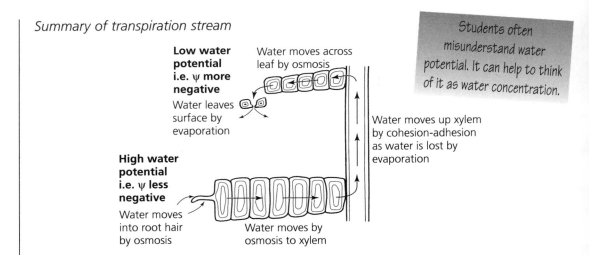

Students often misunderstand water potential. It can help to think of it as water concentration.

Low water potential i.e. ψ more negative

Water moves across leaf by osmosis

Water leaves surface by evaporation

Water moves up xylem by cohesion-adhesion as water is lost by evaporation

High water potential i.e. ψ less negative

Water moves into root hair by osmosis

Water moves by osmosis to xylem

Root pressure is produced by the absorption of salts into the root cells, which increases movement of water by osmosis. Root pressure contributes to water movement in conditions when the transpiration rate is low.

Mineral ions are absorbed from the soil and move into the xylem (sometimes by active transport). Once there they are moved in a mass flow system to where they are needed.

The sugar products of photosynthesis are transported as sucrose in the living phloem vessel. This is known as **translocation**. The mechanism of translocation is not well understood. The way in which sugars are moved in plants is uncertain but flow rates are known to be in the region of 0.2–6mh^{-1}. There is much less phloem than xylem and resistance to flow is a combination of its narrowness and the presence of **sieve tubes**.

Why can't it stay as sucrose?

The sucrose content of most cell sap is about 0.5%, but it is 20–30% in phloem. This suggests active transport is involved in the movement of sucrose into the phloem against such a diffusion gradient. Specialised cells called **transfer cells** seem to be involved in this movement. Movement along the sieve tubes is not fully understood but **mass flow** and the **pressure flow hypothesis** provide models for the mechanism.

In 1930, Munch proposed a theory of movement based on turgor pressure gradients in plants. Differences in turgor pressure cause a mass flow of water through phloem that transports organic substances. This is called the mass-flow hypothesis and the diagram below shows the demonstration model used by Munch.

To begin with, water moves into both containers by osmosis. Since X contains a much more concentrated solute solution than Z, water will move into X more rapidly and so there will be a flow of solution from X to Z. The flow will continue until the concentrations of the solutions in X and Z are the same.

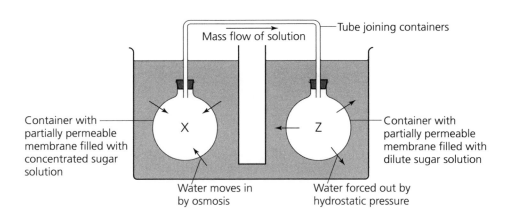

Mass flow of solution

Tube joining containers

Container with partially permeable membrane filled with concentrated sugar solution

Container with partially permeable membrane filled with dilute sugar solution

Water moves in by osmosis

Water forced out by hydrostatic pressure

The model can be applied to a living plant.

X represents the phloem in the leaves where sucrose concentration is high as a result of photosynthesis and the transfer cells are actively loading the sieve tubes. Z represents the area of phloem where the sucrose is unloaded and used by the cells. Water can move into the phloem by osmosis at any point with its return route to the cells being through the xylem. Unlike the physical model, the flow can be continuous because sucrose is continually being added at one end and removed at the other.

Mass flow in plant tissues

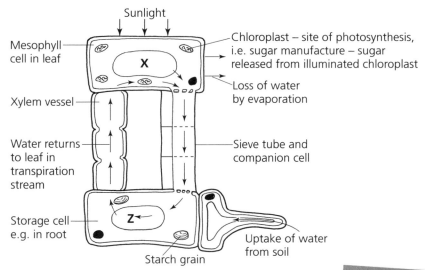

Make a copy of the diagram and mark on how the water potential in the cells changes.

The opening and closing of stomata

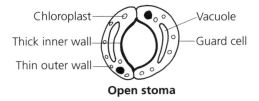

Open stoma

Stoma or 'little mouth' is the opening. The cells are guard cells.

Closed stoma
(stoma is never completely closed)

Stomatal opening and closing depends on changes in turgor of the guard cells.

If water flows into the guard cells by osmosis, their turgor increases and they expand, but they do not expand uniformly in all directions – the relatively inelastic inner wall makes them bend and draw away from each other. The result is that the pore opens. If the guard cell loses water, the reverse happens.

It is known that the guard cells increase their turgor by actively accumulating potassium ions (K^+), which lower their water potential and cause the inflow of water by osmosis. This accumulation of potassium ions requires the expenditure of energy. This is provided by the transfer of electrons during photosynthesis, which generates ATP. That is why the guard cells have chloroplasts!

Transport in animals

In the course of an organism's life materials are constantly being moved around the body. In larger animals complex transport systems have evolved to carry out this function. The transport systems take the form of **cardiovascular systems** which comprise a pump, the **heart**, which pumps a carrying medium, the **blood**, through a series of tubes, the **blood vessels**.

Component of blood	Features
Plasma	Main component of blood. Straw-coloured liquid. Contains soluble molecules, e.g. proteins, hormones, etc.
Red blood cells	Biconcave discs – no nucleus. Contain red pigment – haemoglobin (carries oxygen). Formed in bone marrow. No nucleus – live about 120 days. About 5 million per mm^3 of blood.
White blood cells	Larger than red blood cells. About 7,000 per mm^3 of blood.
Neutrophils (engulf bacteria)	(Or phagocytes) constitute 70% of the total number of white cells. Can squeeze through capillary walls.
Eosinophils (act like antihistamine)	Represent 1.5% of total white cells but numbers increase in allergic conditions such as hayfever. Thought to have anti-histamine properties. Number present is under the control of hormones produced by the adrenal cortex.
Basophils (produce heparin)	Represent 0.5% of white cell population and produce heparin and histamine.
Monocytes (engulf bacteria)	Represent 4% of the white cell population and have a beam-shaped nucleus. Are actively phagocytic and able to migrate from bloodstream to infected areas in a similar way to neutrophils.
Lymphocytes (produce antibodies)	Constitute 24% of the white cell population, are produced in the thymus gland and lymph tissue. Possess very small quantity of cytoplasm. Major function is to cause or mediate immune response e.g. antibody production. Life span from days to years.
Platelets	Tiny fragments of cells involved in blood clotting. About 0.25 million per mm^3 of blood.

Neutrophils, Eosinophils and Basophils are known collectively as **granulocytes**.

Monocytes and Lymphocytes are known collectively as **agranulocytes**.

Check how much detail you need to know in your syllabus.

The functions of blood are:

- transport of digested food materials, e.g. glucose
- transport of excretory materials, e.g. urea
- transport of hormones, e.g. insulin
- maintenance of body temperature
- transport of oxygen
- transport of carbon dioxide
- clotting to prevent blood loss and infection
- provision of immunity (by lymphocytes)
- combating pathogens (by phagocytes)
- acting as a buffer to pH changes.

Formation of intercellular fluid

Intercellular or tissue fluid is formed when blood passes through capillaries – the capillaries are permeable to all components of blood except red cells and plasma proteins.

The solute potential exerted by the plasma proteins is about 3.3 kPa. This is far greater than the solute potential in the tissue fluid – this would suggest that tissue fluid should flow into the blood plasma. However, the blood pressure at the arterial end of the capillary is 4.3 kPa, so fluid passes from the capillary into the tiny spaces between cells to form the **intercellular fluid**. It is through the tissue fluid that exchange of materials between blood and tissues occurs.

The blood cannot afford to constantly lose fluid, so much of it must be returned. This occurs in two ways:

1 At the venous end of the capillary, blood pressure has fallen to 1.6 kPa. This is below the solute potential exerted by the plasma proteins, so there is a net flow of tissue fluid back into the capillary.

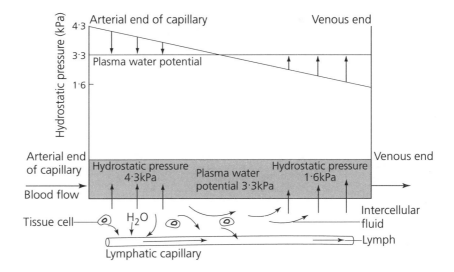

2 The remaining intercellular fluid drains into blind ending lymphatic capillaries and once inside is now called lymph. These capillaries join to form larger lymphatic vessels. Lymph is moved through the vessels by the contraction of the muscles surrounding them, with back flow being prevented by valves.

The lymphatic vessels of the legs join with those from the alimentary canal and empty into the bloodstream in the neck.

Carrying oxygen and carbon dioxide

Oxygen and carbon dioxide carriage

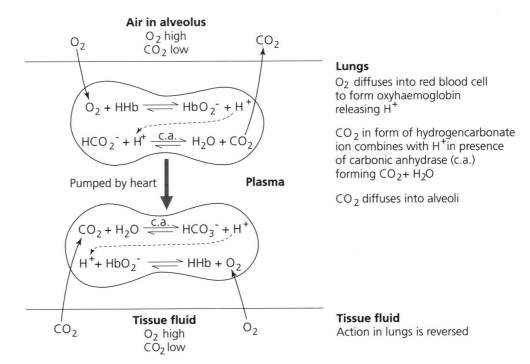

As oxygen binds with a haem group, successive oxygen molecules bind more easily and vice versa. This gives rise to the oxygen-dissociation curve.

N.B. The curve is altered by higher concentrations of carbon dioxide, showing that oxygen is given up more easily. This movement of the curve is called the Bohr Shift – see the diagram below.

Why would oxygen need to be given up more easily at higher levels of CO_2?

Exam questions often ask you to explain a curve which has been drawn under special circumstances. Think about the circumstances before you interpret the curve. It will be more straightforward.

Oxygen dissociation curve and Bohr Shift

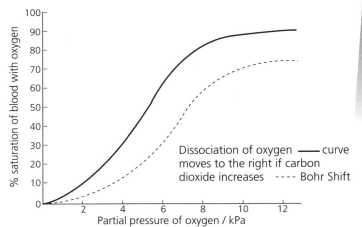

Does temperature affect the dissociation of oxygen? How? Why?

Summary table showing characteristics of blood vessels

If you think you will forget what the different vessels do, remember **arteries always** carry blood **away** from the heart.

Artery	Capillary	Vein
Tunica externa		
Endothelium		
Tunica media		
Tunica externa – thick and made of collagen fibres	Tunica externa – absent	Tunica externa – thin and made of collagen fibres
Tunica media – thick and made of muscle fibres	Tunica media – absent	Tunica media – thin and made of muscle fibres
Endothelium – present	Endothelium – present	Endothelium – present
Transports blood away from the heart at high pressure, and pulsing	Site of exchange between blood and tissues at falling pressure	Transports blood towards heart at low pressure, and flows smoothly

Mini-glossary
endo = inside
externa = external or outside
media = middle
thelium = layer of skin
tunica = tunic or coat

The heart and circulation

Veins have valves. Can you think of an artery that has a valve at its beginning?

The heart is a four chambered structure. The left side pumps blood from the lungs to the body, while the right pumps blood from the body to the lungs. This is called a **double circulation**.

The cardiac cycle is the series of events which constitute the heartbeat:

Make sure you can draw and label a diagram of the mammalian heart.

- atria are relaxed and fill with blood
- blood flows through the valves to fill the ventricles – **diastole**
- both atria contract to force more blood into the ventricles
- ventricles contract forcing blood out of the heart – **systole**.

N.B. The characteristic 'lub-dub' sound of the heartbeat is the sound of blood being forced back against, first, the atrio-ventricular valves and then the semi-lunar valves.

Summary of double circulation

```
           ┌──────────┐
       ┌──→│  Lungs   │──┐
       │   └──────────┘  │
Pulmonary              Pulmonary
artery ↑               vein ↓
       │   ┌──────────┐  │
       │   │  Heart   │  │
       └───│          │──┘
       ┌───│          │──┐
Vena   ↑   └──────────┘  │
cava   │                 ↓
       │   ┌──────────┐  │
       └───│ Tissues  │──┘
           └──────────┘
```

Make a copy of this – it's good revision – and mark on it how blood pressure changes around the circulation.

Remember the left and right side is not **your** left and right, but the left and right of someone facing you, i.e. R← →L.

Control of the heart

Cardiac muscles are said to be myogenic, which means they have an intrinsic rhythm of contraction. Therefore the heart will continue to beat even if its nerve supply is cut.

The rate at which the heart beats, however, is under nervous control. The heart receives two nerves, a **sympathetic** nerve which is part of the sympathetic nervous system, and a branch of the vagus nerve which belongs to the **parasympathetic** nervous system (for further information see the section on the nervous system).

A common question is to ask why the atrio-ventricular node passes the excitation down to the bottom of the ventricles – it delays the impulse so that the contraction has time to pass to both atria. Also it causes contraction from the bottom to push the blood up and out.

Take the opportunity to test yourself on the differences between the sympathetic and parasympathetic systems.

N.B. These do not initiate the beating of the heart but modify the activity of the pacemaker. Experiments have shown that if the sympathetic nerve is stimulated the rate at which the heart beats increases. Stimulation of the vagus nerve slows down the rate. The vagus and sympathetic nerves are therefore antagonistic in their effects.

Excretion and osmoregulation

Excretion is the removal from the body of the waste products of metabolism

Osmoregulation is the process by which the water potential of the blood and tissue fluids of an animal is maintained at a constant level.

Single celled organisms living in fresh water have a specialised contractile vacuole for osmoregulation.

Take the opportunity to check you can explain the principle of osmosis.

Amoeba and the contractile vacuole

Explain what is happening in terms of water potential.

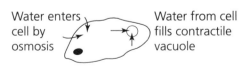

Water enters cell by osmosis   Water from cell fills contractile vacuole

Water discharged into external medium by contractile vacuole 'bursting'

Large multicellular organisms need to control the water potential of tissue fluid which bathes the body's cells, to prevent damage to the cells. Osmoregulation in mammals is brought about essentially by the kidneys, which also act as organs of excretion.

Position of kidneys

Internal structure of kidneys

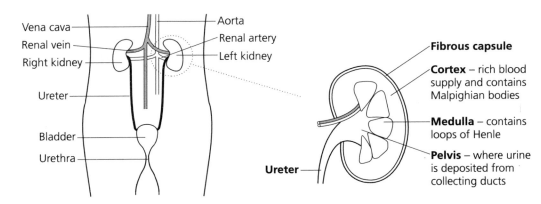

Kidney function can be divided into three:
- ultra-filtration
- selective reabsorption
- tubular secretion

*Draw a simple diagram of a nephron and work on the **three** functions of the kidney. How is nephron structure adapted to these functions?*

*Remember **distal** is more **distant** from the glomerulus.*

Detail of a kidney nephron

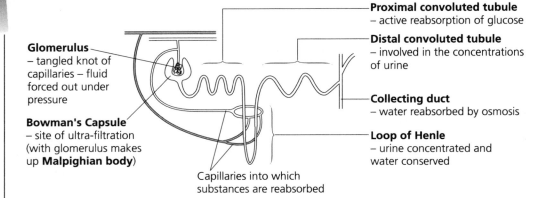

Countercurrent multiplier

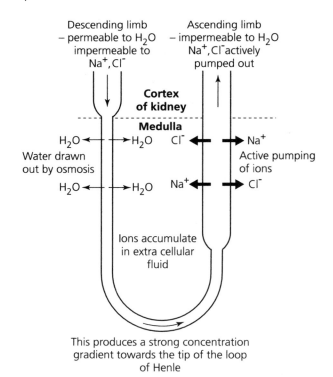

Explain this in terms of feedback mechanism.

The water potential of the blood is maintained in a fairly steady state by the balance of fluid taken into the body and that lost.

Changes in blood concentration are detected by osmoreceptors in the hypothalamus in the brain. This controls the release of a hormone (antidiuretic hormone – ADH) from the pituitary gland, which controls the permeability of the distal convoluted tubule and the collecting duct. If the osmoreceptors in the hypothalamus detect an increased concentration in the blood, ADH is released and the tubule and duct become more permeable. The result is that more water is reabsorbed and the urine becomes concentrated. If blood becomes too dilute, ADH release is inhibited and the tubule and duct walls become impermeable to water, which therefore remains in the urine and is excreted.

Kidney function control

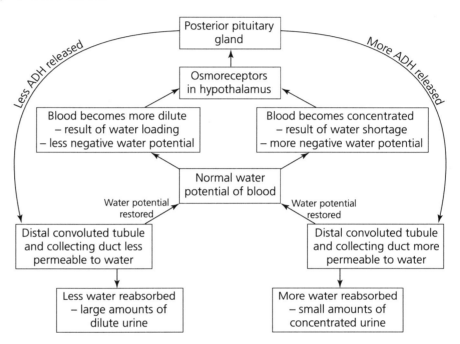

Water balance in plants

See page 11 for more information on water potential.

Water balance is of great importance to plants for the maintenance of turgor and the transport of materials.

Osmoregulation in plants tends towards strategies for preventing water loss.

Plants are classified on the basis of their osmotic habitat:

- **hydrophytes** live in water with, therefore, no obvious shortage
- **mesophytes** live on land and have to balance water loss by transpiration with uptake from the soil
- **xerophytes** live in desert regions so must prevent water loss
- **halophytes** live on seashores and must cope with salty water.

Plants have a variety of methods for preventing water loss:

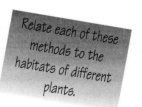

Relate each of these methods to the habitats of different plants.

- waxy cuticle reduces water loss from the leaf surface
- stomata can be situated to reduce loss, e.g. on underside of leaves, with hairs to trap air or sunk into the leaf surface or on leaves of a reduced size such as needles or spines
- succulent plants store water in specialised organs
- roots may be shallow and spreading or very deep
- in adverse conditions leaves are lost and/or seeds/spores produced.

Homeostasis and chemical co-ordination

Can you work out what the word means literally? Split it into two parts: "homeo" "stasis".

Homeostasis is the maintenance of a **steady internal state** and is vital if cells are to function correctly.

Homeostasis in animals requires a high level of control and co-ordination. In particular, any changes in the blood are detected and feedback mechanisms provide a means of controlling this.

A note about feedback mechanisms

Negative feedback means that when something changes, the opposite effect is produced to counteract the change, e.g. an increase in the level of glucose in the blood sets in motion processes which decrease the level. Conversely, a decrease in glucose concentration sets in motion processes which will increase it.

The result is that whatever the direction of the change, the glucose concentration automatically returns to the optimum level. This optimum represents the **norm** or **set point**.

To counteract a change in the concentration of blood glucose, some corrective mechanism must be involved:

Draw this flow diagram again with an example instead of just the headings. How many can you construct?

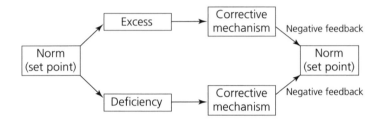

Sometimes a deviation from the norm is not corrected but instead leads to further deviation.

This is known as positive deviation, e.g. amphibian development is controlled by the thyroid hormone, thyroxin. In a tadpole the secretion of thyroxin is kept steady by negative feedback. However, just before the tadpole undergoes metamorphosis, the negative feedback is replaced by positive feedback resulting in a rise in thyroxin levels and this triggers metamorphosis.

Both nervous and hormonal systems are involved in homeostasis and there are a number of examples of homeostatic systems:

Remember: Homeostasis is a good example of the nervous and endocrine systems working together.

- control of water balance or osmoregulation
- control of nitrogenous substances in the blood
- control of blood sugar level
- control of body temperature.

Thermoregulation as a homeostatic mechanism

Temperature is an important environmental factor and the normal temperature range for active life **for most organisms** is between 10° and 35°C.

Animals can be classified as **homoiotherms**, which means they can regulate their body temperature, or **poikilotherms**, which means their body temperature fluctuates largely with that of their environment. More familiar terms are probably 'warm blooded' and 'cold blooded' respectively.

How does the s.a. : vol. ratio change as an organism gets bigger?

The body of a mammal can be thought of as being in two parts, a **central core** where heat is generated and stored, and an **outer shell**, consisting essentially of skin. The amount of heat lost or gained through the shell depends on its nature and thickness as well as the organism's surface area:volume ratio. Normal, healthy, adult humans maintain a body temperature between 36° and 37.5°C. The thermoregulation centre in the hypothalamus in the brain acts like a thermostat, switching on and off heat loss and heat conservation mechanisms as appropriate.

Temperature regulation

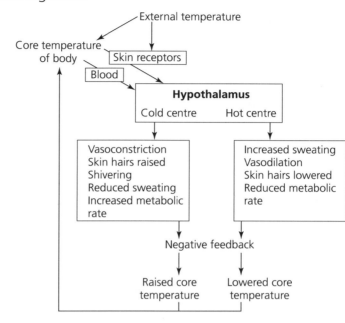

Warm blooded organisms adapt according to conditions, e.g. they are generally larger in cold climates (compared with equivalent organisms in hot climates) and they have thick deposits of fat and/or they may hibernate. Adaptation to hot climates includes having very little fat. Some animals like the camel do not sweat. Both thermoregulation and osmoregulation are examples of homeostasis and exhibit feedback mechanisms. Very often, these mechanisms also involve a system of chemical co-ordination brought about by the **endocrine system**.

The endocrine system

The endocrine system consists of a number of **glands** which manufacture chemical messengers called **hormones**. Hormones are secreted into the blood and transported around the body until certain cells and tissues, known as target organs, respond – usually by bringing about a change in metabolism. Consequently, hormone regulation is generally quite slow and the effects are diffuse. Hormones are released in response to specific stimuli which can be nervous or chemical.

The **pituitary gland** is the most important endocrine gland because it co-ordinates the actions of many of the other glands and is also a link with the nervous system.

Check if you need to know this.

Many hormones, e.g. adrenaline, bind to receptor sites on the target cell membrane and cause the release of a 'second messenger' which initiates a sequence of enzyme mechanisms to produce the appropriate hormonal response. In many cases the second messenger is the nucleotide 3',5'-adenosine monophosphate (cyclic AMP), which is formed by the action of the enzyme adenyl cyclase on ATP following the enzyme's release from the receptor site.

How does ATP become AMP?

Summary of some hormones and their effects

Endocrine structure	Hormone	Target organ	Response
Pituitary gland	Human growth hormone	Body cells	Stimulates protein synthesis and energy release from fats
	Thyroid stimulating hormone	Thyroid gland	Stimulates production of thyroxine
	Adrenocorticotrophic hormone	Adrenal gland	Stimulates release of adrenal cortex hormones
	Follicle stimulating hormone	Ovary	Maturation of follicles
	Luteinising hormone	Ovary	Stimulates ovulation and corpus luteum development
	Prolactin	Mammary glands	Stimulates milk production
	Anti-diuretic hormone	Kidneys	Stimulates water reabsorption in the kidney
Thyroid gland	Thyroxine	Body cells	Increases metabolic rate and growth in infants
Pancreas	Insulin	Liver	Stimulates absorption of glucose
	Glucagon	Liver	Stimulates glycogen breakdown
Adrenal gland	Adrenaline	Body cells	'Flight or fight' response
Ovary	Oestrogen	Genitalia and body Uterus (menstrual cycle)	Development of female sexual characteristics
	Progesterone	Uterus (menstrual cycle and pregnancy)	Development of uterus for implantation
Testis	Testosterone	Male genitalia and body cells	Development of male sexual characteristics

Control of blood sugar level

Blood sugar regulation is another example of homeostasis, but is also an example of a mechanism under hormonal control. Blood sugar level is usually maintained at about 1mg glucose per cm^3 of blood. The liver and pancreas are the main organs responsible for maintaining this level.

The **Islets of Langerhans** are groups of cells in the pancreas. If α **cells** they produce the hormone **glucagon**, and if β **cells** they produce **insulin**.

Can you think of anywhere else where the term 'antagonistic' is used?

Glucagon and insulin are antagonistic, i.e. they have opposite effects.

Antagonistic effects of glucagon and insulin

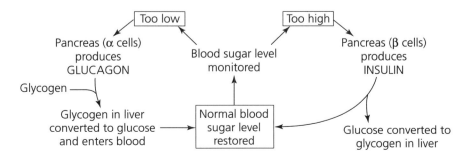

Raised blood sugar levels are known as **hyperglycaemia** and the effect is to stimulate the release of insulin. When blood sugar levels fall – **hypoglycaemia** – the secretion of insulin falls. This interaction of insulin and glucose levels is another example of a feedback loop:

Practice drawing this flow diagram, working on the feedback mechanisms.

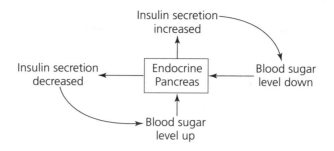

Also sensitive to lower glucose levels are the cells which secrete glucagon. Only liver cells are sensitive to glucagon and they respond in 2 ways:

- increased conversion of glycogen in liver to glucose
- increased rate of glucose formed from amino acids.

Metabolism and the liver

The liver prevents blood glucose levels from fluctuating.

All 6 carbon sugars are converted to glucose and stored as the insoluble polysaccharide, **glycogen** (up to 100g are stored).

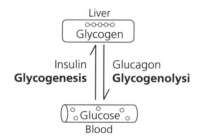

When the demand for glucose has exhausted the glycogen store in the liver, glucose can be synthesised from non-carbohydrate sources. This is called **gluconeogenesis**.

Under hormonal influences, amino acids, glycerol and fatty acids are released from the tissues into the bloodstream. They also stimulate an increased synthesis of liver enzymes which will convert amino acids and glycerol to glucose (fatty acids are converted to acetyl coenzyme A and used directly in Krebs Cycle).

Summary of carbohydrate metabolism

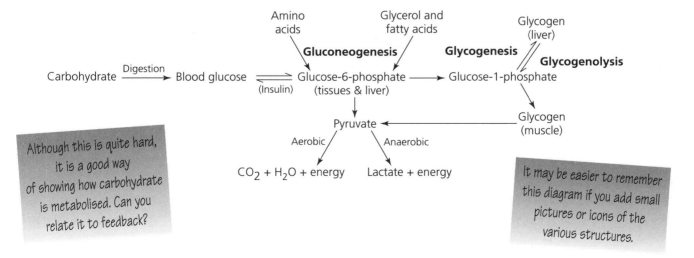

Although this is quite hard, it is a good way of showing how carbohydrate is metabolised. Can you relate it to feedback?

It may be easier to remember this diagram if you add small pictures or icons of the various structures.

The liver also plays an important part in protein metabolism. This occurs by a cyclic reaction known as the **ornithine** cycle:

Ornithine cycle

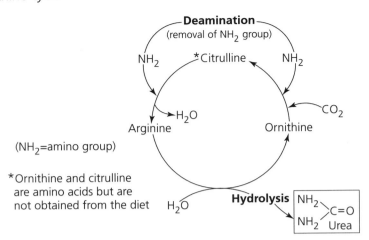

(NH$_2$=amino group)

*Ornithine and citrulline are amino acids but are not obtained from the diet

If they are **not** in the diet where do they come from? Take the opportunity to revise protein synthesis.

Suggested further reading

Eckbert, R. (1988) *Animal Physiology*, W.H.Freeman (0-176-71937)

Flowers, T.J. (1992) *Solute Transport in Plants*, Blackie Academic and Professional (0-216-693220)

Ridge, I. (ed.) (1991) *Plant Physiology*, Hodder & Stoughton/Open University (0-340-53186)

Stewart, M.J. (ed.) (1991) *Animal Physiology*, Hodder & Stoughton/Open University (0-340-53187-8)

Chapter four

Biological basis of co-ordination and behaviour

Glossary

Action potential – temporary and local reversal of the resting potential arising when an axon is stimulated

Antagonistic muscles – muscles which work in direct opposition

Auxin – plant growth substance

Co-ordination – processes by which the activities of an organism are made to function as an integrated whole

Depolarised – reversal of the charge distribution found in the polarised state

Effector – produces a response to stimulation

Nerve – a collection of neurones

Neurone – nerve cell

Polarised – the inside and outside of a neurone membrane carry different charges

Receptor – sense organ that detects change

Reflex action – response to a particular stimulus without conscious control

Reflex arc – the nerve pathway which brings about a reflex action

Refractory period – brief period following the passage of an action potential when an axon is no longer excitable. The **absolute** refractory period is extremely short (1ms), when no matter how strong the stimulus, no response will occur. During the **relative** refractory period the resting potential is gradually restored.

Resting potential – due to the difference in concentration of ions on either side of the neurone membrane (-70mV)

Stimulus – change which is detected and brings about a response

Summation – the addition effect in the post-synaptic membrane of packages of transmitter substance from individual synaptic vesicles

Synapse – gap across which an impulse must pass between one neurone and the next. The synapses can be **adrenergic** (where the transmitter substance is noradrenaline) or **cholinergic** (where the transmitter substance is acetyl choline).

Threshold – the minimum strength of stimulus which will bring about a response

Tropism – plant growth movement in response to a stimulus, e.g. phototropism

How do single-celled organisms and simple multicellular organisms respond to stimuli?

The ability to detect changes (or **stimuli**) and respond to them is called sensitivity. All living things respond to stimuli, whether they are internal or external to the body of the organism. The ability to co-ordinate both internal and external activity is essential to all living organisms. An important requirement of animals, especially, is that this co-ordination is based on a rapid internal mechanism.

Nervous communication

A nervous system is made up of nerve cells or **neurones**, which convey messages from **receptors** that pick up stimuli to **effectors** which bring about a response.

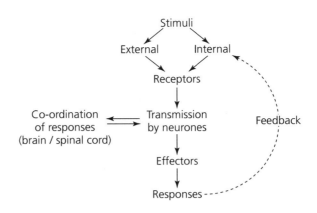

The mammalian nervous system can be subdivided:

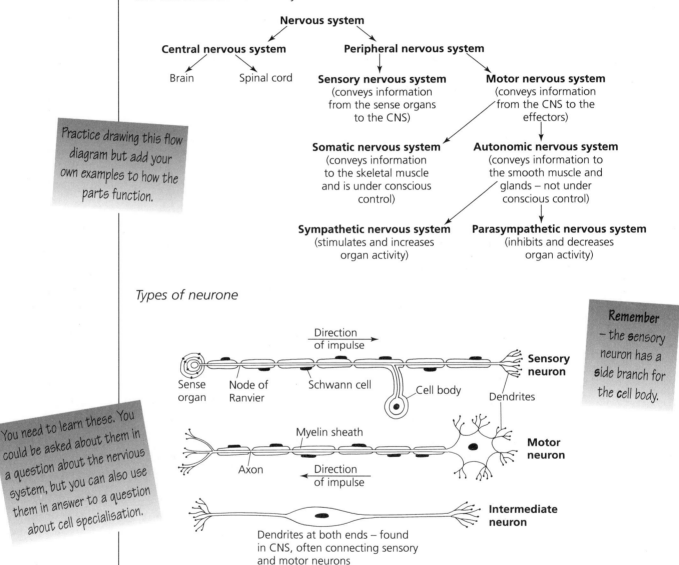

Types of neurone

The transmission of an impulse

The nerve impulse is a tiny electrical event which results from charge differences across the membrane of a neurone. Based on ion movements across the membrane through specialised pores, it involves active transport.

The neurone membrane is impermeable to sodium ions but permeable to potassium ions. Sodium ions are actively pumped out while potassium ions are pumped in (the **sodium-potassium pump**). Potassium ions then diffuse along a concentration gradient.

The result is an imbalance with the inside of the membrane being negatively charged **relative** to the outside (it is described as being **polarised** or having a **resting potential** of -70mV).

Resting potential

Remember – this charge of -70mV is in **comparison** with the outside of the membrane. How is this measured?

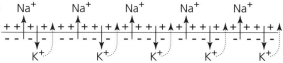

Potassium ions diffuse along a concentration gradient out of the axon so the inside becomes negative relative to the outside

The sodium-potassium pump actively pumps sodium out of and potassium into the axon

Sodium cannot diffuse back into the axon so a concentration of positive ions is built up outside the axon

An impulse or **action potential** is a local, temporary reversal of the resting potential which occurs when an axon is stimulated. When an action potential is set up, **sodium channels** open and sodium rushes in **depolarising** the membrane, i.e. it becomes more positive on the inside than on the outside by about +40mV. The channels then close and the sodium is pumped out again. Permeability to potassium is briefly increased so it diffuses out more quickly than normal. After a brief period known as the **refractory period**, the resting potential is restored.

Make sure you can explain how the sodium pump works.

An action potential

Always remember to mark the direction of the impulse.

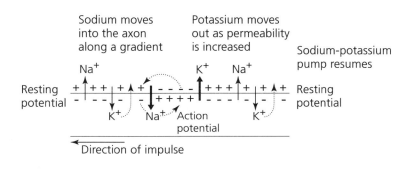

Sodium moves into the axon along a gradient

Potassium moves out as permeability is increased

Sodium-potassium pump resumes

Resting potential

Resting potential

Action potential

Direction of impulse

If you get an exam question about this and/or action potentials, use annotated diagrams – they save time and a lot of writing!

Because of the refractory period, impulses travel 'one way' along an axon – until the resting potential is restored a fibre cannot conduct another impulse.

An action potential explained

Try writing your own explanation for what is happening at stages A–G. Then check to see if you are correct.

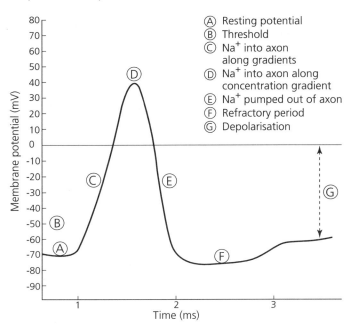

(A) Resting potential
(B) Threshold
(C) Na$^+$ into axon along gradients
(D) Na$^+$ into axon along concentration gradient
(E) Na$^+$ pumped out of axon
(F) Refractory period
(G) Depolarisation

Exam questions often ask for definitions of these terms. If you use the 'graph' you can explain them better.

The refractory period lasts between 5 and 10 ms. The first millisecond is the **absolute refractory period** when no impulse can pass. The remaining time, indicating the 'recovery' of the membrane, is known as the **relative refractory period** during which it becomes increasingly possible to transmit an impulse. An action potential is an **all-or-nothing** response, i.e. it either happens or it does not and is always the same magnitude. A stimulus must be above a certain minimum strength, known as the **threshold**, for an impulse to be generated.

Impulse propagation

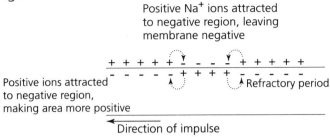

How does this affect the speed of transmission? What happens in non-myelinated fibres? Can you explain how this might affect the organism's response to stimuli?

In myelinated fibres **saltatory conduction** occurs, i.e. the impulse jumps from node of Ranvier to node of Ranvier – the only places where ions can pass freely in and out of the fibre.

Impulse transmission (saltatory conduction) – myelinated fibres can transmit impulses at up to 100ms^{-1}

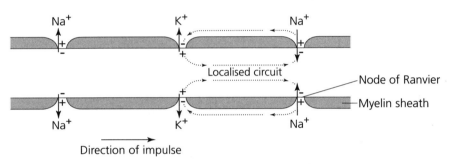

An impulse is transmitted from one neurone to the next at a junction called a synapse.

A synapse

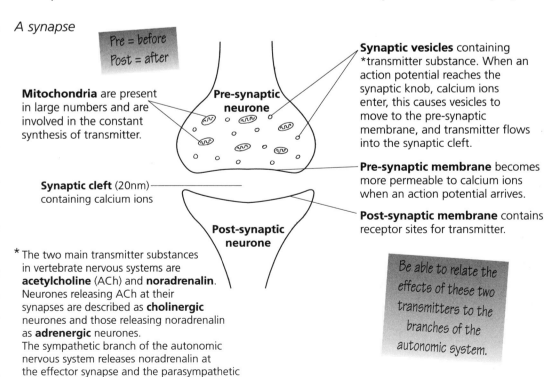

Pre = before
Post = after

Mitochondria are present in large numbers and are involved in the constant synthesis of transmitter.

Pre-synaptic neurone

Synaptic vesicles containing *transmitter substance. When an action potential reaches the synaptic knob, calcium ions enter, this causes vesicles to move to the pre-synaptic membrane, and transmitter flows into the synaptic cleft.

Synaptic cleft (20nm) containing calcium ions

Pre-synaptic membrane becomes more permeable to calcium ions when an action potential arrives.

Post-synaptic neurone

Post-synaptic membrane contains receptor sites for transmitter.

* The two main transmitter substances in vertebrate nervous systems are **acetylcholine** (ACh) and **noradrenalin**. Neurones releasing ACh at their synapses are described as **cholinergic** neurones and those releasing noradrenalin as **adrenergic** neurones.
The sympathetic branch of the autonomic nervous system releases noradrenalin at the effector synapse and the parasympathetic releases acetylcholine.

Be able to relate the effects of these two transmitters to the branches of the autonomic system.

Often a single impulse is not sufficient to develop an action potential in the postsynaptic membrane; instead it takes several impulses to 'build up' an action potential in an effect known as **summation**.

A summation process which involves impulses arriving from several different axons is called **spatial summation**.

Temporal summation occurs when a number of impulses from a single axon follow in rapid succession and have a cumulative effect.

Synaptic transmission can be seriously affected by drugs.

Pictures of where these act can help. Add your own to check your understanding and aid your memory.

Substance	Site of action	Function
LSD	Mammalian brain	Produces hallucinations by mimicking the action of transmitter substances
Nicotine	Post-synaptic membrane	Mimics the action of acetylcholine
Caffeine	Post-synaptic membrane	Mimics the action of acetylcholine
Strychnine	Post-synaptic membrane	Prevents the breakdown of acetylcholine resulting in uncontrollable muscular contractions
Nerve gas	Post-synaptic membrane	Prevents the breakdown of acetylcholine resulting in uncontrollable muscular contractions
Atropine	Parasympathetic post-ganglionic endings	Blocks the action of acetylcholine
Curare	Post-synaptic membrane of neuromuscular junction	Blocks the action of acetylcholine
Analgesics	Mammalian brain	Activates inhibitor synapses making brain neurones resistant to excitation
Tranquillisers	Mammalian brain	Activates inhibitor synapses making brain neurones resistant to excitation

Reception of stimuli

- A large number of actions in organisms are the result of unconscious **reflex actions**.
- These reflex actions, known as **unconditioned reflexes**, are controlled by a nerve pathway known as a **reflex arc**.

A reflex arc

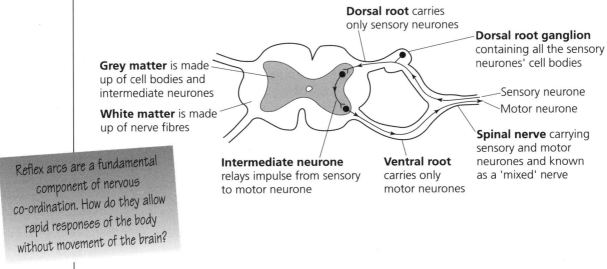

Dorsal root carries only sensory neurones

Dorsal root ganglion containing all the sensory neurones' cell bodies

Grey matter is made up of cell bodies and intermediate neurones

Sensory neurone
Motor neurone

White matter is made up of nerve fibres

Spinal nerve carrying sensory and motor neurones and known as a 'mixed' nerve

Intermediate neurone relays impulse from sensory to motor neurone

Ventral root carries only motor neurones

Reflex arcs are a fundamental component of nervous co-ordination. How do they allow rapid responses of the body without movement of the brain?

N.B. Conditioned reflexes, unlike unconditioned reflexes, are learned and form an important part in the learning processes of behaviour.

Single sensory cells can carry vital information in a reflex arc to bring about a response, but in complex responses the stimulus is detected by groups of sensory cells specialised into regions known as **sense organs**.

Focusing an image on the retina

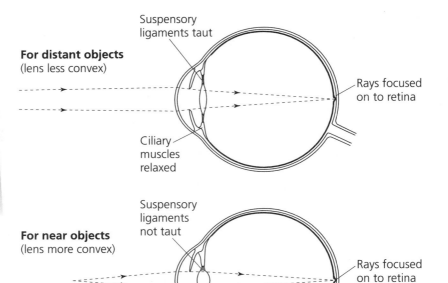

Make sure you can draw and label a diagram of the eye.

It is unusual for an exam question to ask you to draw this, but it could easily ask you to complete a partially drawn version. Practice on unmarked drawings.

Points to remember: most refraction takes place at the cornea; the further away the object, the more parallel the rays of light.

The retina contains photoreceptor cells called **rods** and **cones**. Rods are responsible for black and white vision, and contain a visual pigment called **rhodopsin**. Cones are concentrated at the fovea and are responsible for colour vision; each cone has its own sensory neurone and one of the visual pigments (red, green or blue).

The central nervous system

Complex organisms cannot function efficiently simply by a series of reflex actions – there needs to be a co-ordinating mechanism.

Evolution has provided a branch of the nervous system known as the **central nervous system (CNS)**, which consists of the **brain** and **spinal cord**.

The CNS processes information and co-ordinates responses.

The spinal cord contains **grey matter** made up of neurone cell bodies, and **white matter** containing nerve fibres.

At its anterior end the spinal cord is connected to the brain, which has 3 distinct regions:

- the **forebrain**
- the **midbrain**
- the **hindbrain**.

Make sure you can label a simple diagram of a brain.

Autonomic nervous system

The brain has many co-ordinating functions, such as the control of various body functions via the **autonomic nervous system**.

The autonomic nervous system is concerned with the involuntary responses of the body and can be divided into two distinct and antagonistic parts:

- the **sympathetic** system
- the **parasympathetic** system.

Another of the brain's many co-ordinating functions is to allow the development of complex forms of behaviour and communication.

	Sympathetic system (transmitter substance – noradrenalin; prepares the body for stress)	Parasympathetic system (transmitter substance – acetylcholine; reduces stress)
Heart	Accelerates heart beat	Slows heart beat
Lungs	Dilates bronchioles	Constricts bronchioles
Intestines	Decreases peristalsis	Increases peristalsis
Blood vessels of alimentary canal	Constriction	Dilation
Salivary glands	Reduces secretion	Increases secretion
Adrenal medulla	Secretion of adrenalin and noradrenalin	No effect
Blood vessels of skeletal muscle	Dilation	No effect
Blood vessels in skin	Constriction	No effect
Sweat glands	Increased secretion	No effect

Behaviour

Behaviour can be defined as 'an action in response to a stimulus which modifies the relationship between organism and environment'.

The study of behaviour is called **ethnology**. In **innate** or **species-characteristic** behaviour, an organism has a genetically (unlearned) determined response to a particular stimulus. This type of behaviour can be subdivided:

- **taxes** which are movements of the whole organism in response to an external directional stimulus, e.g. an organism such as an earthworm may move away from light (**negative phototaxis**)
- **kineses** are non-directional movements of an organism to a stimulus; the movement is related to the intensity of the stimulus and not the direction
- **simple reflexes** or rapid responses to stimuli such as potentially dangerous situations
- **instinctive behaviour** consists of sophisticated inborn actions which are species-specific, e.g. courtship, territory defence, etc.

Learned behaviour can also be put into categories:

- **habituation** is the ignoring of a stimulus which is repeated many times, e.g. birds learn to ignore a scarecrow
- **operant learning** is the result of 'trial and error' when an organism may be rewarded or punished; reward results in the behaviour being more likely to be repeated, whereas punishment means it is less likely to be repeated
- **imprinting** is a simple and specialised type of learning which only takes place in very young organisms and involves the identification of a parent

- **exploratory** or **latent learning** happens when an organism explores and familiarises itself with new surroundings
- **insight learning** is based on thought and reasoning, is mainly seen in mammals, especially primates, and is regarded as the highest form of learning.

Effectors and muscle movement

Check with your syllabus that you need to know this. If you do, add some real examples.

Nerves have to communicate with receptors and effectors. Motor nerves need to communicate with muscles to bring about contraction. A special kind of synapse, called a **neuromuscular junction**, occurs where a nerve and muscle fibre meet.

As muscles can only produce a shortening force, i.e. contract, it follows that at least two muscles or sets of muscles must be used to move a bone into one position and back again. Pairs of muscles acting in this way are called **antagonistic** muscles.

Types of movement brought about by pairs of antagonistic muscles

Muscle classification	Type of movement brought about
Flexor	Bends a limb by pulling two skeletal elements towards each other
Extensor	Extends a limb by pulling two skeletal elements away from each other
Adductor	Pulls a limb towards the central long axis of the body
Abductor	Pulls a limb away from the central long axis of the body
Protractor	Pulls distal part of limb forwards
Retractor	Pulls distal part of limb backwards
Rotator	Rotates whole or part of a limb at one of its joints

Neuromuscular junction and simple muscle structure

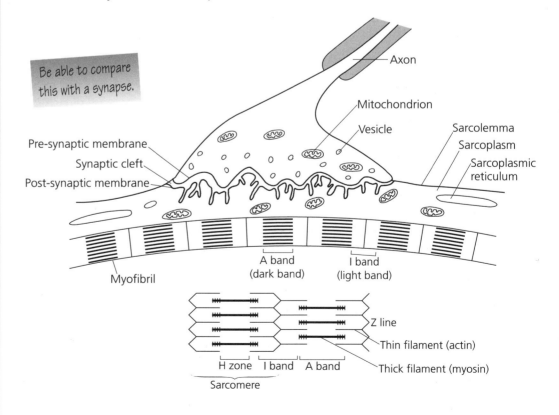

Be able to compare this with a synapse.

Axon
Mitochondrion
Vesicle
Sarcolemma
Sarcoplasm
Sarcoplasmic reticulum
Pre-synaptic membrane
Synaptic cleft
Post-synaptic membrane
A band (dark band)
I band (light band)
Myofibril
Z line
Thin filament (actin)
Thick filament (myosin)
H zone   I band   A band
Sarcomere

Make sure you can relate these points to the stages on the diagrams.

Stages of muscle contraction

1 Nerve impulse travels along a motor neurone and reaches the motor end plate.

2 Acetylcholine is released into the synaptic cleft, diffuses to the sarcolemma and combines with the receptor sites.

3 When threshold is reached, an action potential is created in the muscle fibre.

4 The action potential is conducted to all the microfibrils of the muscle by transverse tubules (T tubules), so spreading throughout the muscle fibre.

5 Calcium ions are released from the sarcoplasmic reticulum and bind to the blocking molecules of the actin filaments, so exposing the binding sites.

6 The 'heads' of the myosin molecules attach to these binding sites on the actin filaments.

7 A 'rowing' action moves the filaments towards the centre of the sarcomeres as the myosin 'heads' move along the binding sites. This results in muscle contraction (the **sliding filament** theory).

8 When nervous stimulation ceases, calcium ions are pumped back into the sarcoplasmic reticulum and the binding sites on the actin filaments are again blocked.

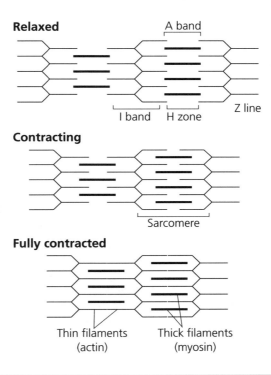

Control systems in plants

Plants are controlled by chemicals which have the effect of controlling growth, i.e. cell division followed by cell expansion. It is at the meristems of plants that most growth takes place. This region is also most sensitive to chemical control. Directional stimuli bring about directional responses known as **tropisms**, e.g. shoots will bend towards the light if illuminated unidirectionally and are described as being **positively phototropic**; roots grow away from light and are called **negatively phototropic**.

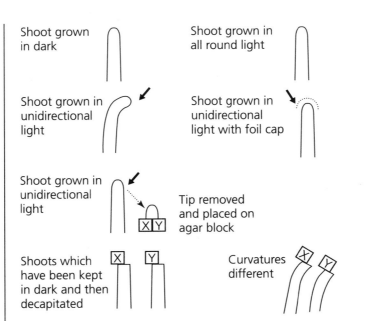

Shoot grown in dark

Shoot grown in all round light

The shoot kept in the dark and the shoot illuminated all-round grow straight up.

Shoot grown in unidirectional light

Shoot grown in unidirectional light with foil cap

The shoot exposed to light from one side grows towards it. Foil cap covers tip and response to directional light is lost, showing the tip is sensitive to light.

Shoot grown in unidirectional light

Tip removed and placed on agar block

Tip removed from shoot exposed to unidirectional light – placed on agar block which is then cut in half. Each half placed on one side of decapitated shoot.

Shoots which have been kept in dark and then decapitated

Curvatures different

Agar block from unilluminated side causes greater growth.

Experiments have shown that the control of tropisms is by plant hormones which are produced at the tips and then transported to the meristems, where they have their effect. The growth substances which bring about phototropism are called **auxins**.

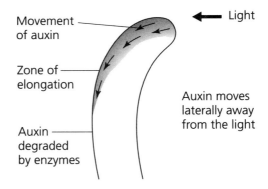

Movement of auxin

Light

Zone of elongation

Auxin moves laterally away from the light

Auxin degraded by enzymes

It is thought that unidirectional light causes auxin to move to the dark side of the shoot. Here it promotes growth, resulting in the shoot tip growing towards the direction of the light.

Another tropism is **geotropism**:
- shoots grow away from gravity – they are **negatively geotropic**
- roots grow towards gravity – they are **positively geotropic**.

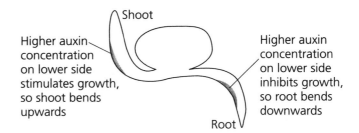

Shoot

Higher auxin concentration on lower side stimulates growth, so shoot bends upwards

Higher auxin concentration on lower side inhibits growth, so root bends downwards

Root

Auxin is thought to be responsible for this, together with a growth inhibitor called **abscisic acid*** and growth promoters called **gibberellins**.

Other effects of auxins:

- pruning apical buds promotes growth of side branches – normally inhibited by the auxin in the apical tip
- growth of adventitious roots from stems is promoted by auxins, so they are used in rooting powders
- auxin helps the setting of fruit and prevents it from falling prematurely
- some auxins are used in weed killers because they can cause abnormal growth.

* Abscisic acid inhibits auxin and gibberellins, and is thought to be involved in leaf and fruit fall because it weakens stems.

Ethene raises respiratory rate and is thought to be involved in the ripening of fruit, leaf fall and release from dormancy of buds and seeds.

Check you need to know this for your syllabus!

Plants respond to light by the action of a photoreceptor called **phytochrome**, a blue-green pigment occurring in 2 interchangeable forms, P_{660} and P_{735}

Red light (rapid conversion)

$$P_{660} \longleftrightarrow P_{735}$$

Far-red (rapid conversion)

Slow conversion in dark

$$P_{660} \xrightarrow[\text{Night}]{\text{Day}} P_{735}$$

Phytochromes control the flowering of plants in response to day length or **photoperiodism**:

- plants which flower when the nights are short are called **long-day plants**
- plants which flower regardless of day/night length are called **day-neutral plants**.

The detection of the length of the photoperiod takes place in the leaves and may be transmitted to the flower buds by a hormone called **florigen**.

Other plant responses include:

- **Nastic responses** are localised responses to generalised stimuli, e.g. flowers which open in response to temperature or sunshine as in daisies; the mechanism by which these are brought about is not known.

- **Thigmotropisms** are responses to touch, e.g. plants which have tendrils that coil around other plants in response to one-sided stimulation as in sweet peas. Some carnivorous plants respond very rapidly, e.g. the Venus flytrap. The mechanism of this rapid response is not fully understood although it is thought to be electrical activity in the cells.
    The slower responses are probably due to similar mechanisms to auxin-induced differential growth.

Suggested further reading

Krebs, N.R. & Davies, B. (1992) *An introduction to behaviour ecology*, Blackwell Scientific Publications (0-632-03546-3)

Manning, A. & Dawkins, M.S. (1992) *An introduction to animal behaviour* (4th edn), Cambridge University Press (0-512-42792-4)

Ridley, M. (1995) *Animal Behaviour* (2nd edn), Blackwell (0-86542-390-3)

Coren, S. & Ward, L. (1989) *Sensation and Perception*, Harcourt, Brace, Jovanovich

Wilkie, D.R. (1976) *Muscle, Studies in Biology Series* No.11, Arnold

McNeil Alexander, R. (1992) *The Human Machine,* Natural History Museum

McNeil Alexander, R. (1992) *Exploring Biometrics: Animals in Motion*, Scientific American Library

Diversity of organisms

Glossary

Analogue – features shared by unrelated organisms which have the same function although their basic structure is quite different, e.g. wings

Binomial – a two-part name

Cladistics – a method of classification based on shared characteristics, which are assumed to indicate common ancestry

Homologue – any structure or biochemical feature which is shared between two or more organisms by virtue of a common ancestral link, e.g. a five-fingered limb

Phylogeny – common ancestry

Taxonomy – study of classification of living organisms

There is an enormous diversity and huge number of living things on the Earth. Estimates of the number of different species have been in the order of up to 100 million.

Diversity of living things

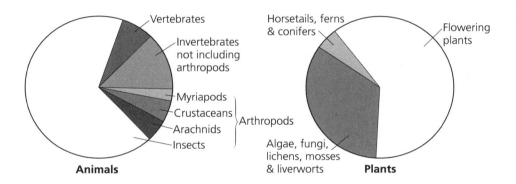

Animals

Plants

Classification and taxonomy

The study of classification is called **taxonomy** or **systematics**.

Biological classification is often described as **phylogenetic**, which means it reflects the evolution of organisms.

One practical way of attempting to classify organisms is to put them into groups based on visible features. However, different stages in life cycles can look very different and this can cause confusion.

A classification system gives an agreed name to an organism and it is the **binomial system** that is internationally agreed.

In the binomial system, devised by Swedish botanist Linnaeus, every organism has a scientific Latin name made up of two parts. The first word is a noun, written with a capital letter, denoting the **genus** of the organism. The second word is an adjective, written with a small letter, denoting the **species** of the organism.

e.g.

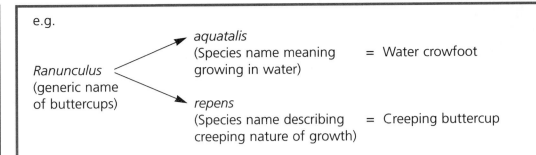

Ranunculus
(generic name
of buttercups)

aquatalis
(Species name meaning
growing in water) = Water crowfoot

repens
(Species name describing
creeping nature of growth) = Creeping buttercup

A **species** represents the first level of classification. Members of the same species have common ancestry, closely resemble one another and are capable of interbreeding to produce fertile offspring.

The name of a genus (the generic name) is shared with related species which are sufficiently similar to be grouped together. It can include any number of species from one upwards.

.

Classification is described as being **hierarchical**, with species placed in groups which in turn are placed in larger groups:

- each successive group contains more and more different kinds of organism
- named organisms are placed in groups on the basis of shared, observed features
- different but similar species with many features in common share the same genus.

The general name for a grouping is a **taxon** and hence the science of classification is called taxonomy, with the lowest taxon being a species and the highest a kingdom:

Species – group of organisms capable of interbreeding to produce fertile offspring
Genus – group of similar/closely related species
Family – group of apparently related genera
Order – group of apparently related families
Class – grouping of orders within a phylum
Phylum – organisms constructed on a similar plan
Kingdom – largest and most inclusive group,
    e.g. animals, plants, etc.

An acronym is a good way of remember the order, e.g. Some Girls Find Offering Chocolates Perfectly Kind.

For example:

Kingdom – Animalia
Phylum – Vertebrata
Class – Mammalia
Order – Primate
Family – Hominidia
Genus – *Homo*
Species – *sapiens* = Modern man

A major issue in classification is how best to divide the living world into kingdoms. Until recently this has been on the basis of a 4-kingdom system:

- Monera / Prokaryotes
- Plants
- Fungi
- Animals.

More recently a 5-kingdom system has been suggested:

- Monera /Prokaryotes
- Protoctista
- Plants
- Fungi
- Animals.

A very brief summary of the 5-kingdom system

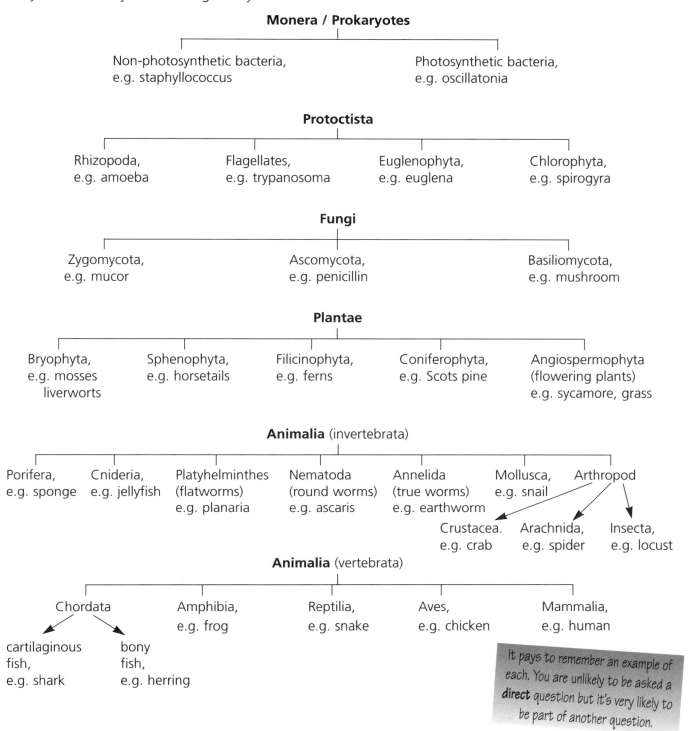

Monera / Prokaryotes

Non-photosynthetic bacteria,
e.g. staphyllococcus

Photosynthetic bacteria,
e.g. oscillatonia

Protoctista

Rhizopoda,
e.g. amoeba

Flagellates,
e.g. trypanosoma

Euglenophyta,
e.g. euglena

Chlorophyta,
e.g. spirogyra

Fungi

Zygomycota,
e.g. mucor

Ascomycota,
e.g. penicillin

Basiliomycota,
e.g. mushroom

Plantae

Bryophyta,
e.g. mosses
liverworts

Sphenophyta,
e.g. horsetails

Filicinophyta,
e.g. ferns

Coniferophyta,
e.g. Scots pine

Angiospermophyta
(flowering plants)
e.g. sycamore, grass

Animalia (invertebrata)

Porifera,
e.g. sponge

Cnideria,
e.g. jellyfish

Platyhelminthes
(flatworms)
e.g. planaria

Nematoda
(round worms)
e.g. ascaris

Annelida
(true worms)
e.g. earthworm

Mollusca,
e.g. snail

Arthropod

Crustacea.
e.g. crab

Arachnida,
e.g. spider

Insecta,
e.g. locust

Animalia (vertebrata)

Chordata

Amphibia,
e.g. frog

Reptilia,
e.g. snake

Aves,
e.g. chicken

Mammalia,
e.g. human

cartilaginous
fish,
e.g. shark

bony
fish,
e.g. herring

It pays to remember an example of each. You are unlikely to be asked a **direct** question but it's very likely to be part of another question.

Identification of organisms

It is often necessary to be able to identify specimens – while doing field work for example. For this **identification keys** are invaluable. A key usually consists of a series of alternative clues derived from the external features of organisms. These clues are based on lists of similarities and differences in an organism's appearance, e.g. number of legs, number of body segments, presence of wings, etc.

External features like these form the basis of a **single-access key**, i.e. one using contrasting characteristics to divide a group of organisms into two groups depending upon whether they possess the feature or not.

These are called **dichotomous keys**. In such keys characteristics are chosen which can be used to separate groups into smaller and smaller groups until a species is identified.

Examples of keys can be found in most books. Make sure you know how to use one that employs pictures as well as the type that asks questions.

Another method of ordering and grouping is called **cladistics**. This is a method of taxonomy based on constructing groups or **clades** comprising organisms which share a unique **homologue**, i.e. any structural or biochemical feature shared between two or more organisms by virtue of an ancestral link. For example, birds form a clade, sharing the unique homologue of feathers, mammals form a clade, sharing the unique homologue of mammary glands and suckling their young.

However, birds and mammals both belong to a bigger clade, the vertebrates, because they both possess backbones.

The relationship between clades is expressed in a branching diagram called a **cladogram**:

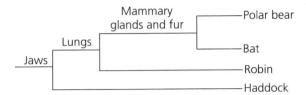

The joints of the branching system represent shared homologues and are called **nodes**.

The cladogram is read from left to right and it can be seen that the groups or clades become smaller. As they do so, the more homologous characteristics they have in common, i.e.
- the polar bear and bat have four features in common (jaws, lungs, fur, mammary glands)
- the polar bear and robin have two features in common (jaws, lungs)
- the polar bear and haddock have only one feature in common (jaws).

The validity of this method of classification depends upon the assumption that if two organisms share a homologue they must be related.

This being the case, the nodes of a cladogram would represent phylogeny (common ancestry) and the entire branching system will reflect evolutionary history.

Suggested further reading

Barnes, R.S.K., Calow, P. & Olive, P. (1993) *The Invertebrates – a new synthesis*, Blackwell Scientific Publications (0-632-0312-1)

Bell, P.R. (1992) *Green Plants, their Origin and Diversity*, Cambridge University Press (0-521-43875-1)

Forey, P.L. (1992) *Cladistics*, Systematics Association Publication Number 10

Jeffrey, C. (1992) *An Introduction to Plant Taxonomy* (2nd edn), Cambridge University Press (0-521-24542-7)

Kershaw, D.R. (1983) *Animal Diversity*, University Tutorial press (07231-084710-1)

Margulis, L. & Schwartz, K.V. (1988) *Five Kingdoms – an illustrated guide to the phyla on earth*, W.H.Freeman (0-7167-1912-6)

Monger, G. & Sangster, M. (1988) *Systematics and Classification*, Longman

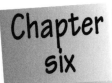

Chapter six

Ecology and environmental physiology

Glossary

Abiotic – the abiotic factors affecting a population are the non-living elements of its environment

Biome – one of the world's major ecosystems, e.g. grassland, forest, desert, etc.

Biosphere – the life-supporting seas, soils and atmosphere of the planet Earth and the organisms which live upon it

Biotic – the biotic factors are all the living elements of the environment of a plant or animal

Carrying capacity – the point of stabilisation or zero growth rate is the maximum carrying capacity of the given environment for the organism concerned

Climax community — the final stable and self-perpetuating community which is in equilibrium with its environment

Commensalism – a way of life where one partner benefits and the other is unaffected by it

Community – the total number of populations of animals and plants living in a habitat at any one time

Decomposer – bacteria and fungi which break down the remains of animals and plants and return the mineral nutrients to the soil

Density-dependent factor – growth rate depends on the numbers present in the population

Density-independent factor – growth rate is not tied to population density

Detritivore – fragments of decomposing material are called **detritus** and the many small animals that feed on these, contributing to their breakdown, are called detritivores

Ecosystem – all organisms inhabiting a particular habitat together with their non-living environment

Habitat – the place where an organism lives, e.g. rocky pool, deciduous woodland, etc.

Mean – often referred to as the average, it is found by adding together all the values and dividing by the number of values

Median – the value which, if the observations are placed in rank order, will divide the distribution into two halves (the middle)

Mode – the most common category or value of a variable – it occurs most often

Mutualism – a way of life involving two or more organisms where the association benefits all partners

Niche – the role of an organism in a community and its way of life – it can be broken down into specific elements, e.g. the food niche or the habitat niche

Null hypothesis – any difference between data sets is purely due to chance

Parasitism – a way of life which benefits one partner (the parasite) and harms the other (the host)

Population – a group of organisms, all of the same species, living together in a particular habitat

Symbiosis – the living together of two or more organisms in close association with each other

Succession – an example of living things (largely plants) altering the abiotic environment is called **primary succession**; successions that occur on soils that have already been formed but have suddenly lost their community are referred to as **secondary successions**

Environmental biology or **ecology** is the study of relationships between living things and their environment.

The organism and its environment

The **biosphere** is the inhabited part of the Earth and extends from the bottom of the oceans to the upper atmosphere

An **ecosystem** is a natural unit of **biotic** (living) and **abiotic** (non-living) components through which energy flows and nutrients cycle.

The biosphere is the largest ecosystem but it can be divided into smaller ones known as **biomes**, e.g. forest, grassland, desert and oceans.

An ecosystem can be divided into various components:
- a **habitat** which is the place where an organism lives, e.g. freshwater pond or rocky shore
- a **microhabitat** is a small part of a habitat, e.g. a tree in a woodland habitat
- a **niche** describes how an organism lives or its role in the community.

Within an ecosystem abiotic or physical factors can influence the living organisms, for example:

- **climatic factors** such as light, water availability, wind and temperature
- **edaphic factors**, i.e. the soil, its texture, acidity, moisture and nutrient content
- **topographic factors** such as north or south facing and angle of slope.

How the environment affects organisms

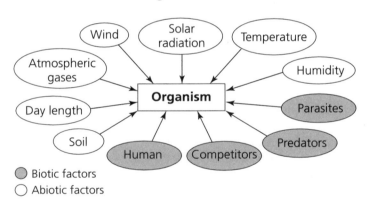

- ○ Biotic factors
- ○ Abiotic factors

Summary of an ecosystem

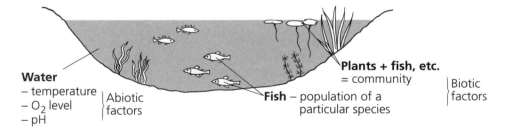

The amounts of inorganic materials on the Earth are limited and are recycled through ecosystems.

This recycling takes place in two phases:
- a **biotic phase** when the materials are incorporated into the tissues of living organisms
- an **abiotic phase** when the inorganic materials are returned to the non-living part of the ecosystem.

Definitions of these terms are often asked for and examiners complain about the confusion – try remembering in alphabetical order because it's also the order of size – biosphere, ecosystem, habitat, niche.

Biotic – with biology i.e. living things Abiotic – without biology i.e. non-living.

Four major cycles are considered.

Work them out – don't try to learn them.

Carbon cycle

- Its abiotic phase is carbon dioxide in the air and dissolved in water.
- Photosynthesis fixes the carbon into organic compounds in plants, which may then be eaten by animals.
- Respiration, death and decay recycle the carbon.

Carbon cycle

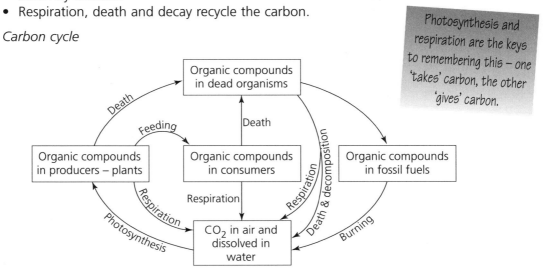

Photosynthesis and respiration are the keys to remembering this – one 'takes' carbon, the other 'gives' carbon.

Nitrogen cycle

- The abiotic phase is nitrogen in the atmosphere and nitrates in soil and water.
- Atmospheric nitrogen is fixed into soil nitrates by **nitrogen-fixing bacteria**.
- These nitrates are taken in by plants to be formed into organic compounds.
- Animals eat the plants.
- Death and decay result in nitrogen compounds in the soil. These are sometimes acted upon by **denitrifying bacteria**, returning nitrogen to the atmosphere.

Nitrogen cycle

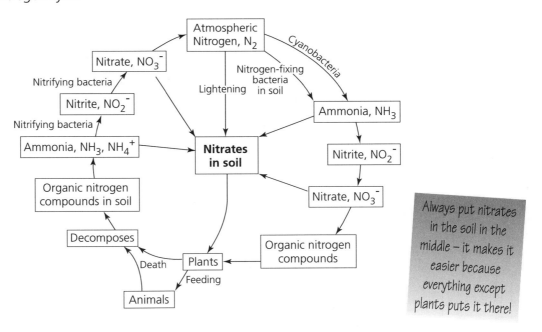

Always put nitrates in the soil in the middle – it makes it easier because everything except plants puts it there!

Water cycle

- The abiotic phase is water in oceans, rivers, etc., plus water vapour in the air.
- Condensation provides water on the land, which may be taken in by living organisms.
- Transpiration in plants returns water to the air.

Water cycle

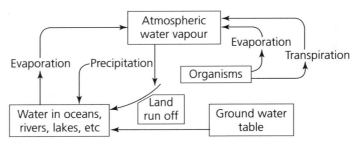

Remember – what goes up must come down!

Phosphorus cycle

- Abiotic phase is in the form of dissolved phosphates in water and soil. Phosphates may enter sediments which become rocks, which in turn may become the basis of fertilisers.
- Plants take in inorganic phosphates, converting them to organic phosphates.
- Animals eat the plants.
- Death and decomposition return phosphates to the soil.

Phosphorus cycle

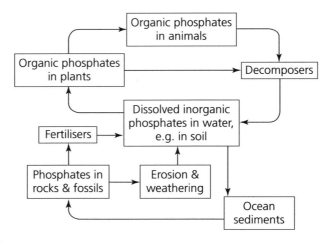

Feeding relationships

Plants are autotrophic, i.e. they can make their own food by photosynthesis.

Animals are heterotrophic, i.e. they need their food 'ready made' and consequently are either directly or indirectly dependent on plants.

So, living things can be studied in terms of their feeding relationships:

- the photosynthetic plants are the **producers**
- the animals are the **consumers**.

Consumers can be:

- **herbivores** which eat plants
- **carnivores** which eat other animals
- **omnivores** which eat both plants and animals
- **detritivores** which feed on dead plants and animals
- **decomposers** which cause the decay of dead material.

The feeding relationships within an ecosystem are, at the simplest level, shown by a **food chain**.

Food chains always start with a producer or autotroph.

The herbivores in a ecosystem are the **primary consumers** with subsequent organisms being **secondary**, **tertiary**, etc., consumers.

These feeding levels are known as **trophic levels**, for example:

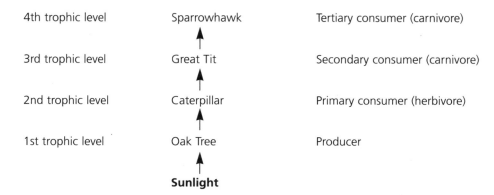

4th trophic level	Sparrowhawk	Tertiary consumer (carnivore)
3rd trophic level	Great Tit	Secondary consumer (carnivore)
2nd trophic level	Caterpillar	Primary consumer (herbivore)
1st trophic level	Oak Tree	Producer
	Sunlight	

Linear food chains, as above, are very simple ways of representing actual feeding relationships.

An ecosystem is actually a complex network of interrelated food chains called a **food web**, for example:

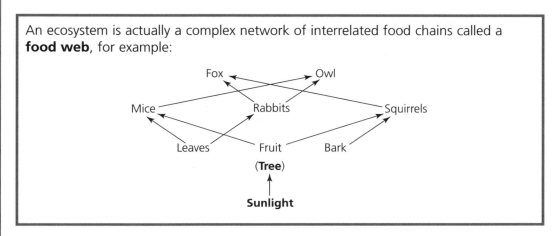

A feature not indicated in a simple food web is the number of organisms occupying each trophic level, e.g. herbivores tend to be less numerous than the plants they eat, and predators tend to be fewer and larger than their prey.

Often, therefore, there is a decrease in numbers from producers to primary consumers, and from primary consumers to secondary consumers, and so on.

This can be shown in a **pyramid of numbers**:

Remember – pyramids of number can be any shape!

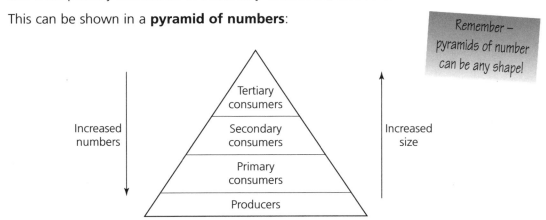

This representation is also flawed because it cannot be applied to all ecosystems, e.g. a **single** large tree can support a **very large number** of smaller herbivores.

A **pyramid of biomass** goes some way to solve this, because it represents the mass in grams of dry material per square metre (gm^{-2}) in an ecosystem, rather than the number of organisms per square metre.

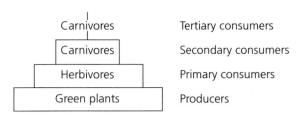

However, there are problems with this representation also. Biomass is measured at a particular time and numbers, and therefore biomass may vary at different times.

Also, it may not look like a pyramid, for example:

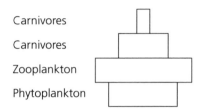

If a continuous sampling were done it would be seen that the phytoplankton reproduces more rapidly than the zooplankton. Therefore, although the population of phytoplankton at any one time (known as the standing crop) is smaller than the zooplankton, the turnover of phytoplankton is much higher and so the biomass over a period of time is greater.

A pyramid made up of samplings taken over time is a **pyramid of energy** and this gives the most accurate picture of energy flow in an ecosystem.

Reliable comparisons between different organisms and trophic levels can be made on the basis of **productivity**, i.e. the mass of new organic material formed in grams of dry mass per square metre per year ($gm^{-2} yr^{-1}$).

As the energy values of proteins, fats and carbohydrates are different, the measurement of productivity is standardised by converting dry mass to its energy equivalent in $kJm^{-2}yr^{-1}$, and this means a pyramid of energy can now be constructed.

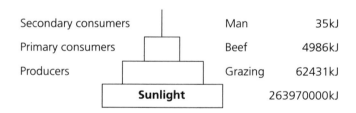

This now clearly shows the **energy transfer and wastage** between trophic levels.

This energy transfer leads to an accumulation of biomass as new plant and animal material is produced. Within an ecosystem there are two kinds of production:
• **primary production** which is the result of photosynthesis
• **secondary production** which is processing and storage by consumers.

Since all consumers ultimately depend on producers in an ecosystem, the rate at which plants convert the Sun's energy must determine the energy flow within the ecosystem.

Only a small percentage of the Sun's energy is transferred into plant material and not all the products of photosynthesis are available to consumers. For example, only about 10% of material is passed on. The remainder is used in, for example, respiration and metabolism.

Suggested efficiencies (very approximate):
- photosynthesis (light to plant) <1–2%
- primary consumer (plant to herbivore) 5–20%
- secondary consumer (herbivore to carnivore) 5–20%
- tertiary consumer (carnivore to carnivore) 5–20%.

The net rate of production can be expressed:
Net primary production (NPP) = Gross primary production – respiration and metabolism

Relatively little of the Sun's energy reaches the Earth. Of that, little is used to produce plant material and increasingly small amounts find their way into the individuals in an ecosystem.

Summary of energy exchanges in an organism

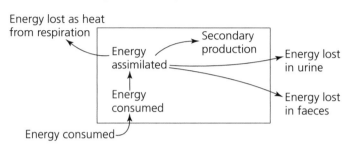

Try writing some percentages on here – it's a good summary to remember.

Summary of energy flow in an ecosystem

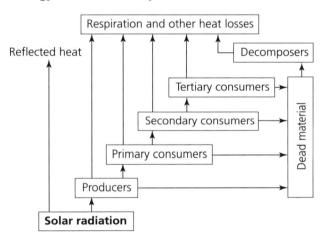

Associations between organisms

The effects of organisms on each other are known as biotic factors. There are times when biotic factors have more effect than abiotic factors on the distribution of organisms.

Symbiosis occurs when two or more organisms of different species live in close association with each other. The nature of that association determines if it is:
- parasitism
- commensalism, or
- mutualism.

Very often names will help you remember what they do if you try to 'understand' the word.

> **Parasitism** is an association in which one organism, the **parasite** lives on or in another organism, the **host**. The parasite depends on the host for food and the host derives *no* benefit from the association. Parasites which live within their host are called endoparasites, e.g. tapeworm, while those living on the surface of the host are called ectoparasites, e.g. mistletoe. Parasites usually kill their hosts eventually.

Commensalism is an association of organisms in which one organism benefits but the other, the host, is neither harmed nor derives any benefits.

Mutualism is an association of organisms in which both organisms benefit, e.g lichens are associations of fungi and algae which have actually resulted in the development of a new species.

In any community, organisms must compete for resources, e.g. plants compete for light and animals compete for food.

Competition between organisms of the same species is known as **intraspecific competition** and forms the basis of natural selection. Competition between individuals of different species is called **interspecific competition**.

A **predator** is an organism that feeds on another species, is normally larger than its prey, and tends to kill before it eats.

N.B. This actually also includes the eating of plants by herbivores, although it is usually called **grazing**. The numbers of predators are limited by the abundance of prey and relationships usually show what are known as **predator–prey oscillations**.

Predator–prey relationships: lynx and snowshoe hare over 100 years.

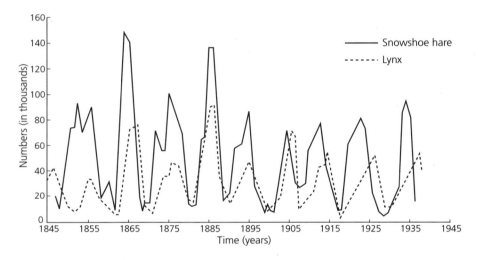

The populations of predator and prey are regulated by **negative feedback**, which maintains the number of organisms at levels the environment can support:

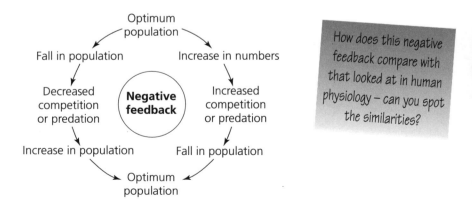

How does this negative feedback compare with that looked at in human physiology – can you spot the similarities?

Populations and communities

All the individuals of a single species in a particular area are known collectively as a **population**, while a **community** is the total number of all the populations in that area.

Population growth can be seen by an S-shaped curve with distinctive stages or phases.

Typical growth curve for a population

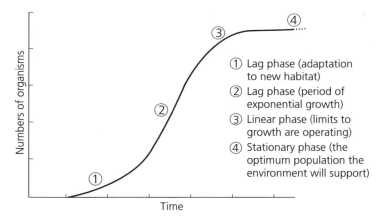

Compare this with the growth curve for bacteria – how are they different and why?

① Lag phase (adaptation to new habitat)
② Lag phase (period of exponential growth)
③ Linear phase (limits to growth are operating)
④ Stationary phase (the optimum population the environment will support)

The size of a population of a species is usually expressed as the number of individuals per unit area and is known as the **population density**.

Population density can vary for a number of reasons:
• birth rate rarely remains constant
• death rates are variable
• mobile organisms move away (emigration)
• new individuals arrive (immigration)
• effects of other species, e.g. competition (known as density-dependent factors)
• changes in one or more abiotic factors, e.g. temperature (known as density-independent factors).

A total count of all the organisms of a population is called a **census**. In ecology this is usually impossible so **sampling** is required (see later).

Populations cannot continue to grow exponentially, but reach a maximum size known as the **carrying capacity** for the particular environment in which the population occurs.

Population growth is affected by **environmental resistance**, the effects of which can vary with the species:
• lack of food or water
• lack of light
• lack of oxygen
• predators and parasites
• disease
• lack of shelter
• accumulation of toxic waste
• stress
• weather and other catastrophes.

Try to think of examples for these – an exam answer always looks better with named examples.

Where do populations and communities come from?

New populations and communities can arise in completely new situations, e.g. a river delta or cooled volcanic lava.

Initially there may be no soil, but minerals may be present and the first plants, known as **pioneer plants**, eventually appear, e.g. algae, lichens, mosses. Because of these plants, organic matter begins to form a simple soil which can now support tiny herbaceous plants. The organic material continues to build as a result of decomposition of organic materials, and larger herbaceous plants and shrubs become established. This sequence of events is known as **primary succession**.

A succession beginning in dry land is called a **xerosere** and one that begins in water, a **hydrosere**. The stages in a succession are called **seral stages**, with the whole succession being a **sere** and the final outcome a **climax community**.

Primary succession

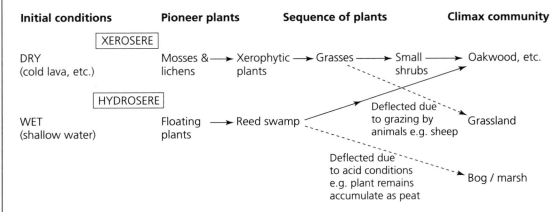

A climax community can remain relatively unchanged for long periods, but only as long as climate and other abiotic and biotic factors allow, e.g. a woodland is vulnerable to winds or disease such as Dutch elm disease.

Sometimes successions occur on soils already formed but which have 'lost' their community and these are known as **secondary successions**, e.g. fire may destroy all the surface layer of a woodland but soon a succession of plants appears on the fire site and the burnt patch quickly re-establishes itself.

Ecological techniques

Synecology is the study of a group of organisms associated together as a community. It involves the analysis of both biotic and abiotic aspects of the community and the environment.

A wide variety of methods is used to sample, estimate and count the abundance, density and effects of organisms, and to estimate the importance of climatic and edaphic factors.

Abiotic factors

Climatic factors have important effects on ecosystems, e.g. annual rainfall, seasonal variations in light level, temperature variations.

Smaller variations constitute the **microclimate** of the different parts of a habitat. The most important microclimate features are:

- **temperature** of the soil, water and air, which can be measured easily using a thermometer to show up variations, such as differences in air temperature between a forest and open countryside

- **light** – its intensity, wavelength and duration are important for organisms which photosynthesise; these can be measured using a range of light-sensitive monitoring equipment

- **wind speed and direction** is measured using a wind-speed meter and by keeping records over a period of time, allowing comparisons to be made

- **humidity** is a measure of the amount of water vapour in the air, which affects water loss by evaporation and transpiration, and is measured using a hygrometer

Edaphic factors are the composition and properties of soil which contribute to the development of plant communities. A series of techniques are used to determine soil properties.

- **Soil sedimentation test** A sample of soil is shaken with water and allowed to settle; different sized particles settle at different rates.

Soil sedimentation

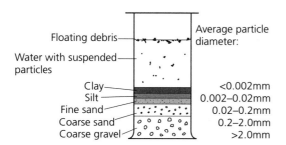

Floating debris

Water with suspended particles

Clay
Silt
Fine sand
Coarse sand
Coarse gravel

Average particle diameter:

<0.002mm
0.002–0.02mm
0.02–0.2mm
0.2–2.0mm
>2.0mm

Water content

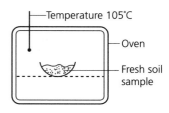

Temperature 105°C

Oven

Fresh soil sample

Organic matter content

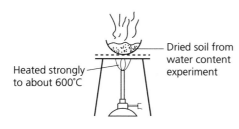

Dried soil from water content experiment

Heated strongly to about 600°C

Air content

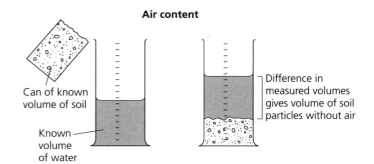

Can of known volume of soil

Known volume of water

Difference in measured volumes gives volume of soil particles without air

Drainage and water retention

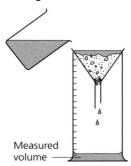

Measured volume

NB. Drainage rate can be calculated by measuring the time taken for a known volume of water to pass through the soil.

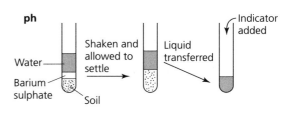

pH

Water

Barium sulphate

Soil

Shaken and allowed to settle

Liquid transferred

Indicator added

- **Estimation of water content** The percentage water content is determined by drying a sample of soil to constant mass

$$\text{water content (\%)} = \frac{(\text{fresh mass} - \text{dry mass}) \times 100}{\text{fresh mass}}$$

- **Estimation of organic matter** A sample of soil from the water content investigation is heated strongly to burn off the organic matter

$$\text{organic content (\%)} = \frac{(\text{dry mass} - \text{mass after heating}) \times 100}{\text{fresh mass}}$$

- **Estimation of air content** Using a can of known volume, a sample of fresh soil is added to a known volume of water and shaken to expel the air

$$\text{air content (\%)} = \frac{\text{volume of can} - \text{difference in measured volumes of water} \times 100}{\text{volume of can}}$$

- **Estimation of drainage/water retention** The time taken for a known volume of water to percolate through a sample of soil gives the drainage rate, and a measure of the volume percolated gives the retention of the soil.

- **Determination of pH** Soil is mixed with barium sulphate and distilled water, shaken and left to settle. The liquid is then tested with universal indicator – the colour gives an indication of the pH.

Biotic factors

Measuring these involves counting the number of organisms which make up a community. The method used depends upon the organisms being investigated.

- **Micro-organisms** can be suspended in water and counted under a microscope using a **haemocytometer** which allows an estimate to be made of the number of organisms present in the original sample. Algae, single-celled fungi and bacteria[*] can be counted in this way.

  [*]An alternative way of estimating bacteria populations is to grow them on agar plates and count colonies.

- **Small invertebrates** such as insects and mites can be counted after they have been extracted from a soil sample using a **Tullgren funnel**, while nematode worms can be extracted using a **Baermann funnel**:

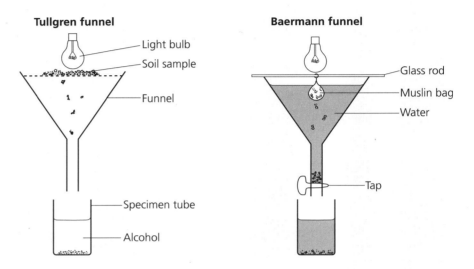

- **Capture-recapture method** is a useful way of estimating the population numbers of fast moving and/or flying organisms. For example:
  - collect 100 insects such as ladybirds (f1)
  - mark them with cellulose paint and release
  - some time later make a new collection (f2)
  - count the number of marked individuals in this collection (f3)
  - an estimate of the total number of ladybirds (N) can now be calculated:

$$N = \frac{f_1 \times f_2}{f_3}$$

If $f_1 = 100$
$f_2 = 75$
$f_3 = 53$
Then,

$$N = \frac{100 \times 75}{53}$$
$$= \underline{142}$$

You probably won't be expected to remember this but you must know how to use it.

Capture-recapture can only give accurate results if:
- the population is confined within a geographical area, e.g. pond, woodland, etc
- organisms are evenly distributed within the area
- marking does not affect survival or probability of recapture
- marked individuals mix randomly with the rest of the population.

Any slow moving or stationary organisms, e.g. plants, are sampled differently. It is often difficult to make a total count of plants, so a statistical approach has to be adopted and some of the methods of sampling are as follows.

- **Quadrat methods** use a square frame usually of 1/2m dimensions called a **quadrat**, which is dropped at random in the area under investigation. Where it lands, the quadrat encloses a known area which can then be analysed to reveal information about the species present.

The simplest analysis involves recording the presence or absence of a particular species. After a minimum of 100 throws of the quadrat, a **percentage frequency** can be calculated:

% frequency of species = $\dfrac{\text{number of quadrats containing the species}}{\text{total number of quadrats thrown}}$

Although quick it gives no information about the number of individuals per square metre (the **density**) which can only be arrived at by counting the number of individuals of each species inside the quadrat throws and taking an average.

- **Transects** measure changes in vegetation due to succession along an environmental gradient, e.g. a seashore. The simplest form is a **line transect** which is a length of string across the area under investigation. The species present at fixed intervals along the string are noted. By stretching two strings parallel to each other and sampling between, a **belt transect** can be carried out.

- The **point quadrat** was developed to give more reliable results for low growing vegetation. It consists of a free standing frame with a row of 10 sliding pins, the points of which can be lowered down onto the vegetation.

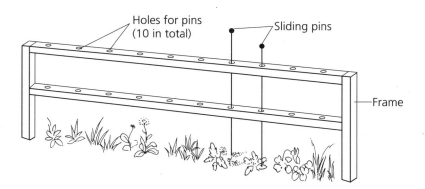

As each point is lowered to the ground, a record is kept of the number of times out of ten (because there are 10 pins) that each species of plant is touched. These are called **hits**. The number of point quadrats used in a sampling area is usually ten and this gives the number of units per 100 points (10 pins x 10 samples).

Percentage cover = $\dfrac{\text{Hits}}{\text{Hits + Misses}}$ x 100

For example,
- If 15 point quadrats are used in a sampling area and the number of hits for a particular species is 20.
- The total number of hits and misses is 150 (since each quadrat has 10 pins)
- Therefore, the percentage cover = $\dfrac{20}{150}$ x 100

    = 13.3%

Recording findings

Once data has been collected it needs to be summarised and put into a more manageable form. Graphs make it easier to understand the relationships between large numbers of figures since they provide a 'picture' of what is happening. There are a number of different graphs and it is important to know which to use.

Line graphs

These represent a quick and simple means of showing the relationship between two variables. Interpretation of a line graph is relatively easy since any trends are quickly seen.

Exam questions often require you to plot a graph – make sure to get the axes the right way round. Check that the question doesn't give you instructions about them.

Bar charts

These are often used for presenting discontinuous data for the purpose of comparison. Discontinuous data may be grouped into one category or another, unlike continuous data which has no such distinct boundaries, e.g. height, weight, etc. The data are represented by drawing bars proportional in height to the value they represent. This form of graph is particularly good for comparing group data such as the density of successive samples of a population.

If a question requires you to plot a bar chart it will tell you. If not, a line graph will be expected.

Histograms

These are similar to bar charts but in this case continuous data is recorded. Unlike the bar chart, the proportion of the data is depicted not only by the height of the column but also by its area. If more than one set of data is to be displayed on the same axes it is often best to superimpose the graphs.

If a question asks you to interpret a histogram, don't forget that not just the height of the columns but the area covered must be discussed as well.

Kite diagrams

Kite diagrams are used most frequently when displaying data from transects. The transect line is laid out and estimations of plant and animal abundance made, e.g. percentage cover of plant species. The data is then plotted as symmetrical line graphs either side of a base line.

Scatter graphs

Sometimes it is preferable to plot points as dots without joining them up with a line. This may be the case when investigating the relationship between two variables and there are data from many locations. In a scatter graph the dependent variable is the y axis and the independent variable, x axis. The pattern of scatter of the points demonstrates any relationship between the two variables. The graph may show a tendency to form a band, sloping at an angle to the horizontal, highlighting a positive correlation between the two variables, i.e. if one goes up, the other goes up. There may seem to be two separate groupings which could correspond to two separate species. Lines of 'best fit' can be drawn to highlight the trend of the dots.

Make sure you understand that a 'line of best fit' is used to show a trend not to join points.

Pie graphs

These differ from other graphs because there are no x and y axes. They are used to display proportional data, the whole quantity being divided into its component parts, for example:

A circle is drawn which represents the whole quantity and it is then divided into segments each of which is proportional to the size of a component.

Things to remember:
- draw a circle proportional in area to the total quantity you want to represent
- compile a table of the values which will form the segments and convert to percentages
- calculate the angle which corresponds to this percentage of 360
- draw a vertical line from the centre of the circle to the top of its circumference

- draw in the segments, measuring the angles already calculated – work clockwise from the vertical.

Ecology and statistics

There are various ways of summarising data.

The mean

This is often known as the average and is found by adding up all the values and dividing the total by the number of values. The mean is shown by the symbol x. The mean is the most commonly used data summary and it can be used for further analysis.

EXAMPLE:

Data: Values – 3, 4, 4, 4, 6, 6, 9 Number of values – 7

Calculation: $\dfrac{3 + 4 + 4 + 4 + 6 + 6 + 9}{7} = \dfrac{36}{7} = 5.1$

$x = 5.1$

The mode

The mode is the most frequently occurring event, e.g. in a series of numbers the mode is the most frequently occurring number.

EXAMPLE :

Data: 3, 4, 4, 4, 6, 6, 9

Mode: most frequently occurring number = 4

If a question involves any or all of these it will probably be in the form of a set of results for analysis.

The median

The median is the central value in a series of ranked values.

If the series has an even number of values then the median is the mid-point between the two centrally placed values.

EXAMPLE :

Data: 3, 4, 4, 4, 6, 6, 9

Median: central value = 4

Data: 3, 3, 4, 6, 8, 9

Median: mid-point between two central values = 5

The mean, mode and median provide a summary value for a set of data, but they can be misleading because they give little indication of the spread of data around the summary value.

Standard deviation – spread around the mean

The standard deviation (σ) enables us to measure the spread of data around its mean value. This is important because the greater the range or spread of the data, the less useful is the mean as a summary of it.

How to calculate the standard deviation

1 Draw a table of the values (x) and then square each of them (x^2).

2 Add these values, i.e. add all the values for x (Σx) and all the values for x^2 (Σx).

3 Find the mean of all the values of x ($\bar{x}$) and square it ($\bar{x}^2$)

4 Calculate the formula:

$$\sigma = \sqrt{\frac{\Sigma x^2 - \bar{x}^2}{n}}$$

Where σ = standard deviation
$\sqrt{}$ = square root of
Σ = the sum of
n = the number of values
x = the mean of the values

The higher the standard deviation, the greater the spread of data around the mean. The standard deviation is the best measure of this spread because it takes into account *all* the values under consideration.

EXAMPLE:
Shell height of dog whelks on a rocky shore

Size classes (mm)	x	x^2
26	5	25
27	3	9
28	10	100
29	13	169
30	11	121
31	12	144
32	17	289
33	6	36
34	6	36
35	3	9
36	2	4
	$\Sigma x = 88$	$\Sigma x^2 = 942$

mean $\bar{x} = \dfrac{88}{11} = 8$ and $\bar{x}^2 = 64$

n = 11

$$\sigma = \sqrt{\frac{\Sigma x^2 - \bar{x}^2}{n}}$$

$$= \sqrt{\frac{942 - 64}{11}}$$

$$= 4.7$$

This figure of 4.7 is the standard deviation of the data from the mean. This means the true mean lies within the range 8 ± 4.7.

Tests of significance

Many pieces of work involve collecting two or more sets of data from different locations with a view to comparing them, e.g. numbers of woodlice in dry and damp conditions. Significance tests are used to find out whether the difference between two or more sets of sample data are truly significant or purely due to chance. If as a result of the tests we determine that the occurrence was due to chance, then we conclude one of two things:

Either

1 The relationship is not significant and there is little point in looking for an explanation

Or

2 The sample is too small and that if a bigger sample were taken then the result of the significance test might change, i.e. the relationship becomes more certain.

The chi-squared (χ^2) test

To use this test the data must have the following characteristics.
- It must be frequency data counted in each of several categories.
- The total number of observations should be more than 20.
- The expected frequency in any one category should be no less than 5.
- The observations should not be such that one influences another.

> Check with your syllabus that you need to know this.

How to calculate χ^2

1 State that any difference between the sample data sets was purely by chance (this is known as the **null hypothesis**).

2 Calculate the 'expected frequency', i.e. the values **expected** to occur if the null hypothesis is correct.

3 Calculate the formula:

$$\chi^2 = \frac{\Sigma(O - E)^2}{E}$$

> Most exam questions will supply you with the formula.

Where χ^2 = chi-squared figure
Σ = sum of
O = observed frequency
E = expected frequency

4 Calculate the 'degrees of freedom' which is one less than the total number of categories i.e. $df = n - 1$

where df = degrees of freedom
n = number of test categories

5 Using the calculated value for chi squared and the degrees of freedom, use a graph or table to read off the probability that the data frequencies being tested could have occurred by chance.

EXAMPLE:
The number of smooth periwinkles found amongst the fronds of four species of seaweed were as follows:

Seaweed	No. of smooth periwinkles
Spiral wrack	5
Bladder wrack	16
Egg wrack	23
Serrated wrack	56

> Chi-square questions don't have to be ecological – they can be genetics questions.

Question: Are these results a true reflection of the distribution of the smooth periwinkle on seaweed fronds or are they simply the result of chance?

The null hypothesis is that the distribution was purely by chance.

If the species of seaweed had no effect on the density of smooth periwinkles then an equal number of animals would be found on the fronds of each species of seaweed, i.e. the 100 individuals divided by the number of seaweed species (4). This means the expected frequency of periwinkles for each species of seaweed is 25 or E = 25.

Seaweed species	Observed frequency (O)	Expected frequency (E)	$\dfrac{(O - E)^2}{E}$
Spiral wrack	5	25	$\dfrac{(5 - 25)^2}{25} = 16.0$
Bladder wrack	16	25	$\dfrac{(16 - 25)^2}{25} = 3.2$
Egg wrack	23	25	$\dfrac{(23 - 25)^2}{25} = 0.2$
Serrated wrack	56	25	$\dfrac{(56 - 25)^2}{25} = 38.4$
			$\Sigma = 57.8$

$$\chi^2 = \frac{\Sigma(O - E)^2}{E} = 57.8$$

Degrees of freedom (df) = n − 1

Where n = number of categories, i.e. species of seaweed = 4

Therefore, df = n − 1

$\qquad\qquad$ = 4 − 1

$\qquad\qquad$ = 3

If you get a chi-square question you will be supplied with a table or part of a table in order to read off the degrees of freedom.

From a graph or tables read off the degrees of freedom (3) against the chi-square value (57.8). The resulting point indicates that the probability that the data could be due to pure chance is less than 1 in 1,000. This means that the evidence is strongly against the null hypothesis and it must be discarded.

Conservation, pollution and human ecology

The earliest humans were hunter-gatherers whose population was kept under control by predators, disease and a limited food supply. This type of society eventually gave way to a semi-agricultural one with the population consuming all the food it produced – known as **subsistence agriculture**.

Eventually, as a result of improved agricultural techniques, societies developed which began to produce surplus food. Continued improvements in all aspects of society, such as a falling death rate, stimulated a massive growth in the human population which continues today. There is a maximum number of people the Earth can support – steps need to be taken to conserve the Earth's ecosystems and prevent a major catastrophe.

Conservation

Conservation involves the management of the Earth's resources in order to restore and maintain a balance between the needs of humans and other species. This can take place from an international down to an individual level.

It should be considered for a number of reasons:

- **ethical reasons** – man has no right to destroy ecosystems and allow species to become extinct
- **practical reasons** – by maintaining the rain forests, for example, the greenhouse effect is reduced, and by conserving fish stocks a food supply is maintained.

There are many ways in which attempts are being made to conserve and support both the human population and other species.

Examiners complain that candidates get too 'emotional' and often waffle on this topic – stick to the facts.

Food production

- The use of **fertilisers** can dramatically increase crop yields but the manufacture of fertiliser is energy intensive, relying heavily on fossil fuels. Excessive use of fertilisers can lead to river and lake pollution.

- Since almost half the world's food production is lost because of various pest species, **pest control** is used, but the chemicals are expensive to produce and can cause environmental damage.

- **Plant breeding** of crops such as wheat, rice and maize has contributed to dramatic increases in crop yields, but this can have only a limited impact in poorer countries because of problems with droughts, floods, etc., and the need for expensive fertilisers.

- By **farming a new species** to be exploited as a resource, food production can be increased, e.g. ostrich farming, but this can be expensive to start and has to deal with the problems of 'acceptance'.

- **Factory farming** is another method of increasing production but is a contentious issue. Less contentious is the culture of single-celled organisms to produce high-protein food.

Soil erosion

- This is often a major problem when land is cleared of covering vegetation, but it can be prevented:

- **Contour farming** is ploughing and cultivating across a slope rather than up and down it.

- **Reduced ploughing** is either shallow ploughing or planting crops through existing vegetation.

- **Wind breaks** such as hedgerows prevent the loss of surface soil.

- **Gully reclamation** by building small dams across rivers behind which silt is trapped eventually creating soil.

> Flow diagrams always help to put the process into perspective and make it easier to understand and recall.

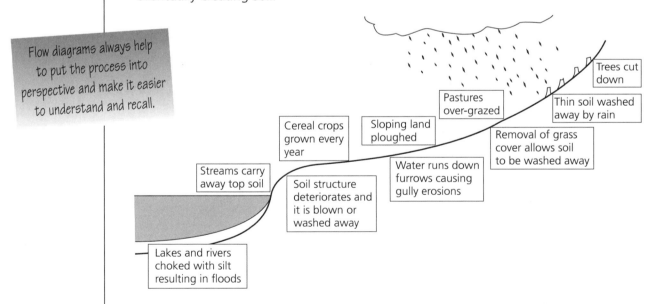

Irrigation, pest control and pesticides

- These are other measures used in an attempt to address the demands of an ever-increasing human population.

- Care must be taken that other species are not ignored as measures are adopted to conserve the human population.

- For example, important measures include saving the tropical rain forests, international agreements regarding whaling, captive breeding programmes to help save endangered species, etc.

- Until a few thousand years ago humans were just one species among many others, but today human activity seems to have some influence everywhere in the world.

Pollution

Pollution is the damaging release by humans of materials or energy (known as **pollutants**) into the environment.

Pollutants can be put into two categories:

- **biodegradable pollutants** which are broken down quite quickly into harmless substances, e.g. sewage

- non-biodegradable pollutants which cannot be broken down easily, therefore accumulate in the environment and as a result are potentially more dangerous, e.g. some plastics, heavy metals.

Air pollution

Pollutant	Major Source	Effects
carbon monoxide	natural methane oxidation, car exhausts, burning fossil fuels	combines with haemoglobin to produce asphyxia
carbon dioxide	burning fossil fuels	contributes to greenhouse effect
hydrocarbons	burning petrol and oil	cause cancer
sulphur dioxide	coal-fired power stations	acid rain, lung damage
nitrogen oxides	car exhausts	asphyxia, smog – asthma
dust particles	car exhausts, industrial chimneys	lung damage
radioactive isotopes	nuclear accidents, nuclear waste	cause cancer
chlorofluorocarbons	coolant, aerosol sprays	destruction of ozone layer

Combustion of fossil fuels produces increasing amounts of carbon dioxide and other gases which diffuse into the upper atmosphere.

The gas layer produced has increased by almost 20% in the last 150 years and causes the enhanced **greenhouse effect** – sunlight is allowed to pass through the layer but the reflected heat from the Earth is trapped by that layer.

Examiners still complain that candidates confuse the greenhouse effect and the hole in the ozone layer – do you understand the difference?

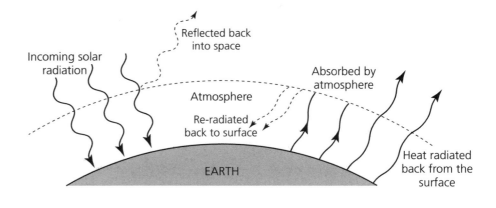

- The greenhouse effect is an example of **global pollution**, i.e. it affects the whole planet, as does the ongoing damage to the ozone layer.

- ozone is constantly being formed in the stratosphere by the reaction between UV radiation and oxygen; but it is being broken down at a faster rate as a result of human activities. For example, chlorine acts as a catalyst in the conversion of ozone to oxygen and is released into the atmosphere from compounds called **chlorofluorocarbons** (CFCs) which are used in cooling systems, foam plastic manufacture and as propellants in aerosol sprays.

Acid rain

This is the result of a variety of processes which together lead to acidic gases being deposited in the atmosphere:

- sulphur dioxide from power stations and nitrogen oxides from car exhausts react with water and oxygen in the atmosphere to produce acids

- unpolluted rain is slightly acidic because of dissolved carbon dioxide (pH 5.6), but polluted rain is very acidic (pH 3–4.2)

- acid rain inhibits tree growth, especially in spruce and pine, which reduces the activity of nitrogen-fixing bacteria, and makes lakes so acidic that no living organism can survive.

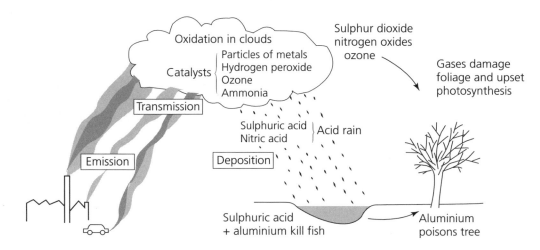

Thermal pollution

This results from the need to generate electricity:

- most power stations are located by water because large amounts of cooling water are needed to re-condense the steam which drives the turbines

- when this cooling water is discharged, into rivers and lakes, for instance, the subsequent rise in temperature stimulates bacterial growth and reduces the amount of oxygen dissolved in the water.

Toxic waste

This includes waste from chemical works, oil refineries, steel plants, paper mills, etc., and is a major source of river pollution. Also the use of landfill sites for dumping waste can pollute water if chemicals, etc., are carried off by drainage water.

Fertilisers and sewage

Both contribute to water/river pollution, either by run-off or the discharge of organic materials into water courses. These organic materials stimulate the growth of decomposing bacteria, which use up most or all of the available oxygen, resulting in the death of other organisms.

How sewage pollutes

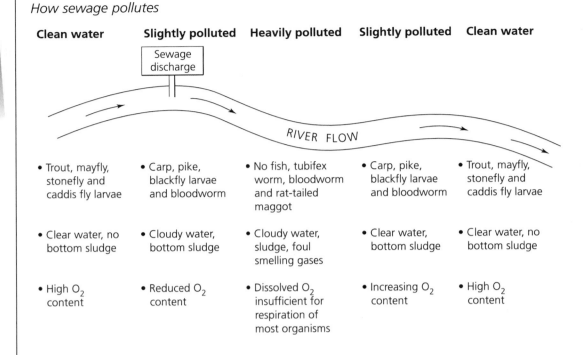

Clean water	Slightly polluted	Heavily polluted	Slightly polluted	Clean water
• Trout, mayfly, stonefly and caddis fly larvae	• Carp, pike, blackfly larvae and bloodworm	• No fish, tubifex worm, bloodworm and rat-tailed maggot	• Carp, pike, blackfly larvae and bloodworm	• Trout, mayfly, stonefly and caddis fly larvae
• Clear water, no bottom sludge	• Cloudy water, bottom sludge	• Cloudy water, sludge, foul smelling gases	• Clear water, bottom sludge	• Clear water, no bottom sludge
• High O_2 content	• Reduced O_2 content	• Dissolved O_2 insufficient for respiration of most organisms	• Increasing O_2 content	• High O_2 content

Suggested further reading

Abraham, W.G. (1989) *Plant Animal Interactions*, McGraw-Hill

Attridge, T.H. (1991) *Light and Plant Responses*, Cambridge University Press (0-521-42748-7)

Chapman, J.L. & Reiss, M.J. (1992) *Ecology – Principles and Applications,* Cambridge University Press (0-521-38951-8)

Gilbertson, D.D., Kent, M. & Pyatt, F.B. (1990) *Practical Ecology*, Unwin Hyman (0-04-445375-2)

Hart, J.W. (1990) *Light and Plant Growth*, Cambridge University Press (0-04-581023-0)

Tivy, J. & O'Hare, G. (1990) *Human Impact on the Ecosystem*, Oliver and Boyd (0-05-003203-8)

Williams, G. (1987) *Techniques and Fieldwork in Ecology*, Bell Hyman (0-7135-2730-7)

Chapter seven

Micro-organisms and biotechnology

Glossary

Adsorption – molecules of one substance attaching to the surface of another substance

Bacteriophage – a virus which attacks a bacterium

Bacterium – prokaryotic unicellular micro-organism belonging to the Kingdom Monera

Batch fermentation – a process in which the product is harvested at the end

Binary fission – division of a cell into two identical 'daughter' cells

Callus – mass of undifferentiated cells which form at the wound site of a plant

Continuous fermentation – a long-term operation in which medium is added as fast as it is used up and the product is harvested

Downstream processing – recovering the product of fermentation from the mixture in the fermenter

Endosmosis – movement of water into a cell by osmosis

Fungus – heterotrophic eukaryotic organism with cell wall containing chitin, usually with a mycelium consisting of thread-like hyphae – many fungi are saprophytic

Generation time – life span of a cell

Immobilisation – act of physically or chemically trapping enzymes or cells onto surfaces or inside gels, etc., the enzyme can be re-used

Parasite – an organism which obtains food from another organism (the host) eventually killing it

Saprophyte – organism which releases enzymes into the environment and then absorbs the simple soluble products of digestion

Symbiosis – the living together in close association of two or more organisms of different species

Total count – a count of all cells, both living and dead

Viable count – a count of living cells only

Virus – consists of nucleic acid surrounded by a protein coat and can only survive and reproduce inside a living cell

Micro-organisms are considered to be those organisms whose size makes them impossible to be seen without a microscope. They include animals like protozoans and plants like algae, together with fungi and bacteria which are not obviously plants or animals.

Micro-organism sizes

Organism	Example	Size
Virus	Tobacco mosaic	300 x 15nm
Alga	Chlamydomonas	30 x 10μm
Fungus	Yeast	12 x 6μm
Bacterium	Typical coccus	1.0μm diameter
Blue-green alga	Chlorogloea	8.0μm diameter
Protozoan	Amoeba	150μm diameter

N.B. 1mm (10^{-3} m) = 1000 micrometers (μm)
$\quad$ 1μm (10^{-6} m) = 1000 nanometers (nm)

Make sure you remember the scale and size of things.

Groups of micro-organisms

Viruses

Viruses are non-cellular particles which lack all the structures characteristic of cells – sometimes they consist of nothing more than nucleic acid and protein. They range in size from 20nm to 300nm.

Viruses can only reproduce themselves inside living cells and are consequently called **obligate parasites**. Once inside the host cell they 'switch off' the host's DNA and use their own DNA or RNA to instruct the cell to make new copies of the virus. Most viruses consist of a protein coat (the capsid) enclosing a nucleic acid, either DNA or RNA.

The capsid is made up of identical repeating sub-units called **capsomeres**. These form highly symmetrical structures which can be crystallised.

Some viruses have an additional **envelope** made of carbohydrate or lipoprotein. A few viruses contain one or two enzymes. The 'fully assembled' unit is called a **virion**.

Three distinct viral shapes have been identified:

Helices

These have conical capsomeres arranged helically in the capsid in which the RNA is embedded, e.g. the tobacco mosaic virus (TMV). The virions look like hollow rods at low magnification. In other helical virions, the helical capsid is coiled and contained in an envelope, e.g. myxoviruses which cause diseases such as mumps, measles and influenza.

Helices

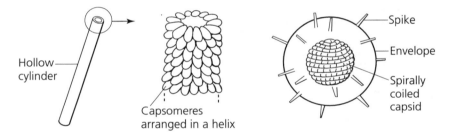

Tobacco mosaic virus　　　　　**Myxovirus**

Be able to draw sketches of the types – it saves writing time.

Polyhedrons

These have the capsomeres arranged into a many-sided shape such as an icosahedron, which has 20 faces, each of which is an equilateral triangle. This group of viruses includes enveloped forms such as the Rubella virus which causes German measles, and the Herpes virus (the cause of cold sores), together with 'naked' forms like the adeno-viruses which cause infections of the tonsils and adenoids. The nucleic acid is found in the capsid, sometimes sandwiched between the capsid and the inner protein coat. The nucleic acid can be DNA or RNA.

Polyhedrons

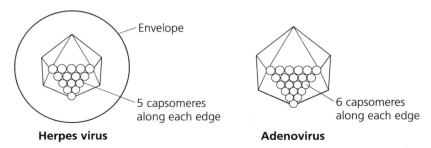

Herpes virus　　　　　**Adenovirus**

Complex viruses

The best known examples are the **bacteriophages** (bacterial viruses), which have an icosahedral head and a helical tail. They also possess a tail collar and a tail plate from which long filaments extend. DNA is found in the head region and comprises a single strand.

Complex virus

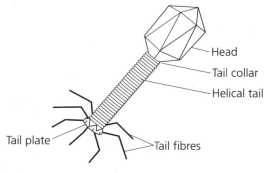

Bacteriophage

Some common human viral diseases

Name of disease	Caused by	Body parts affected	Method of spread
influenza	myxovirus	respiratory passages	droplet infection
common cold	large variety	respiratory passages	droplet infection
smallpox	variola virus	skin	droplet (from wounds)
mumps	paramyxovirus	respiratory and glands	droplet infection
measles	paramyxovirus	respiratory and skin	droplet infection
German measles	Rubella virus	respiratory and skin	droplet infection
poliomyelitis	polio virus	nervous system	droplet or faeces
yellow fever	arbovirus	blood vessels and liver	ticks and mosquitoes

Life cycle of a bacteriophage

1 Phage approaches bacterium and tail fibres fit into receptor sides on the bacterial cell surface

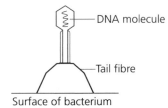

2 Tail fibres bend to anchor to the cell surface; tail contracts forcing a hollow spike into cell and DNA is injected

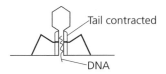

3 Phage DNA codes for production of phage enzymes using ribosomes, etc. of the host

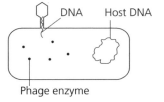

4 Phage inactivates host DNA; phage DNA takes over cell machinery

5 Phage DNA replicates itself and codes for new coat problems

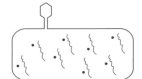

6 New phage particles made by assembly of protein coats

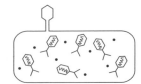

7 Cell bursts and about 200–1,000 phages released to infect more bacteria

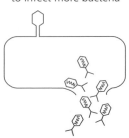

A similar life cycle probably occurs in most viruses, but the penetration process differs in animal and plant and bacterial viruses (a cell wall must be penetrated in plant and bacterial cells). Penetration does not always involve the injection process and protein coats are not always left outside.

Some phages do not replicate inside the host cell. Instead their nucleic acid becomes incorporated into the host DNA and they may remain without influence through many generations, being replicated when the host replicates its DNA. These phages are known as **temperate phages** and an inactive phage is called a **prophage** or a **provirus**.

Fungi

At least 100,000 species of fungi are known to man. They include yeasts and moulds, which can only be seen under a microscope, and mushrooms and toadstools which can reach a substantial size. Many species are useful. The ability of some to recycle nutrients plays an important role in ecosystems. Some can cause serious problems by 'spoiling' products such as food, and some cause death and disease in plants and animals.

Some of the biochemical activities of fungi have long been exploited by man, e.g. in making bread, beer and wine, and in cheese and yoghurt production (see later).

Fungi are relatively simple organisms. They lack chlorophyll which means they cannot photosynthesise. Most fungi live as **saprophytes** on dead organic materials, often in soil. Some grow as **parasites** of plants and animals. Some form **symbiotic associations** with algae to form lichens, and others with the roots of higher plants.

Fungi usually consist of fine, branched threads called **hyphae** which form a tangled mass called the **mycelium**. A hypha can be 0.5µm to 1.0µm in diameter. There may be no cross walls or septa inside the hypha, in which case it is called **coenocytic** or **aseptate**.

Septate hyphae have cross walls which divide the hyphae into compartments similar to cells, with each compartment containing one, two or more nuclei.

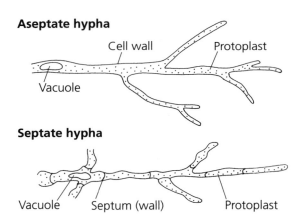

Most fungi reproduce by means of spores, which ensure dispersal and enable organisms to withstand extreme environmental conditions.

Two types of spore exist – asexual which have a dispersal role, and sexual which are produced for survival:

- **Asexual spores** are formed in a number of ways, the simplest is when cells along the hypha become enclosed within a thickened wall and then detach (**chlamydospores**). Another type, the **conidiospore**, is produced in a number of ways, e.g. a budding process forms on distinct hyphae called **conidiophores**. In some fungi asexual spores are formed inside a cell or **sporangium**, and are called **sporangiospores**.

- **Sexual spores** are found as four different kinds and are characteristic of particular sub-divisions: zygospores; oospores; basidiospores; arcospores.

Bacteria

Leeuwenhoek described bacteria in 1683 but it was Pasteur and Koch in the 19th century who advanced the study.

Bacteria are the smallest organisms to have a cellular structure. They occupy many environments in enormous numbers, e.g. 1cm of fresh milk may contain more than 3,000 million.

The following characteristics are typical of bacteria:
- they are generally unicellular, often exist in colonies and some are filamentous
- they are surrounded by a cell wall, which is usually rigid but in some species is flexible
- they usually reproduce by binary fission, with a few species multiplying by budding
- although some bacteria have photosynthetic pigments, most are colourless and feed as saprophytes, parasites, symbionts or commensals
- all bacteria are prokaryotic.

Although some bacteria cause diseases such as TB, pneumonia and diphtheria, most of the 1,500 known species are not pathogenic and some are helpful, e.g. the bacteria of the nitrogen cycle.

Bacteria are classified on the basis of their shape.

Classification of bacterium

Cocci
(spherical)

Staphylococcus

eg. Causes pneumonia, food poisoning

Streptococci

eg. Causes scarlet fever, sore throat

Bacilli
(rod-shaped)

eg. Causes typhoid

eg. Causes anthrax

Spirilla
(spiral-shaped)

eg. Causes syphilis

Vibrios
(comma-shaped)

eg. Causes cholera

Structures *always* found in a bacterium	Structures *sometimes* found in a bacterium
Cell wall	Flagellum
Cell membrane	Capsule/slime layer
Ribosomes	Photosynthetic membranes
Food reserve	Plasmid (extra DNA)
Cytoplasm	Pili or fimbriae
DNA	Mesosome

You should be thinking how this links with the features of prokaryotes – without looking write them down...

Generalised bacterium

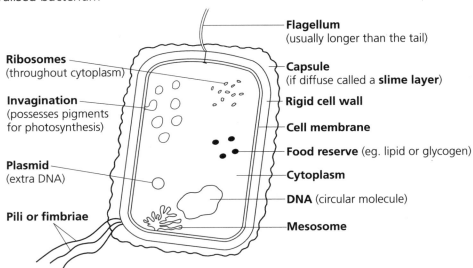

Bacteria possess a cell wall which is relatively rigid. It gives shape to the bacterium and stops bursting should endosmosis occur. The main component of the cell wall is a complex polymer, a **mucopeptide** substance or **murein**.

Bacteria fall into two groups according to their cell wall structure:
- some are stained with Gram's stain and are called **Gram positive**
- some resist the stain and are called **Gram negative** (see later for staining technique).

Many bacteria have a **capsule** around the outside of the cell, consisting of polypeptides and polysaccharides which protect the cell from desiccation.

A **slime layer** has a similar function as the capsule but is more diffuse and less rigid.

The protoplast is the living part of the cell. It is bounded by a cell membrane with a structure similar to that found in eukaryotic cells.

Invaginations of the membrane are called **mesosomes**. They contain a variety of enzymes which catalyse respiration and, in some pathogenic bacteria, produce toxins.

The cytoplasm appears dense and contains ribosomes, storage products such as glycogen and starch and, in photosynthetic species, **chromatophores** containing light-absorbing pigments.

This can be important in recombinant DNA technology.

> Bacterial DNA is a single, circular molecule containing several thousand genes which regulate protein synthesis at the ribosomes, and replicate at fission. Many species of bacteria can move rapidly using **flagella**. A cell may have one or many flagella which may be distributed either over the whole surface or be present at one or both ends.

Fimbriae are much smaller than flagella and enable bacteria to adhere to solids and so remain in contact with a food supply.

Sex fimbriae or **pili** are used to transfer DNA during conjugation.

Bacterial growth

Under favourable conditions bacterial cells grow quickly and soon reach their maximum size. On reaching this size, they reproduce asexually by **binary fission**.

Cell division is preceded by replication of the DNA. While this is being copied it may be held in position by the mesosome.

The mesosome may also be attached to the new cross-walls that are laid down between the daughter cells and play some role in the synthesis of cell wall material.

In the fastest growing bacteria such divisions may occur in 20 minutes – this is known as the **generation time**.

Not all the progeny survive, so the **total** count (living and dead cells) differs from the **viable** count.

	Number of cells
Initial cell	1
First generation	$2 = 2^1$
Second generation	$4 = 2^2$
Third generation	$8 = 2^3$
Fourth generation	$16 = 2^4$
Fifth generation	$32 = 2^5$
nth generation	$N = 2^n$

• A **growth curve** can be plotted of the counts of cells over a period of time:

Growth of a population of E. coli

Hours	No. of viable cells	Total no. of cells
0	20,000	20,000
2	21,900	27,200
4	496,000	540,000
6	5,430,000	6,400,000
8	81,900,000	105,760,000
12	83,400,000	126,300,000
24	80,500,000	127,600,000
36	1,120,000	127,900,000

> An exam question may ask you to plot a graph from a table similar to this – be very careful with the scale for the number of cells. It's easier to use the log of the number. Check with your syllabus to see if you should know how to use log graph paper.

The numbers are so large that the log of the number of bacteria is taken to enable a graph to be plotted:

a – **lag phase** when cell size rather than cell numbers is increasing

b – **log phase** when bacteria multiply rapidly as there is no limiting factor, i.e. there is an ample supply of nutrients and oxygen and all other factors which affect growth are optimal

c – **stationary phase** when the population number remains static, with the 'birth rate' almost balanced by the 'death rate'

d – **death phase** when those individuals unable to compete die, and there is now a marked difference between the viable and total counts. Ultimately, the amount of energy-rich material available becomes nil and no organism can survive in an active state (spores may be produced for survival).

Haemocytometry

A haemocytometer is a special slide that can contain a known volume of a liquid, allowing the total number of cells in the liquid to be determined. Although designed for counting blood cells, the haemocytometer can be used to count any cells that are of a similar size, e.g. yeasts, single-celled algae and immobile protozoa. The haemocytometer slide is about 5 times thicker than an ordinary microscope slide, with a bevelled edge

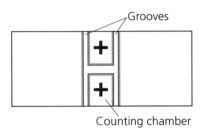

Grooves

Counting chamber

and two or three deep grooves cut across the central portion. A slide with two parallel grooves has only one chamber, while one with three grooves forming an 'H' shape has two chambers on the same slide. This allows for two counts from the same sample without the need to reassemble the slide.

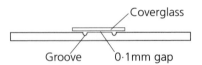

Coverglass

Groove   0·1mm gap

If a cover glass is placed over the grooves and the slide is observed from the side, a small space can be seen below the cover glass between the grooves. When the cover glass is placed correctly this gap measures exactly 0.1mm.

Looking at the slide from above, the area between the grooves is lightly silvered, and by reflecting a light off this surface a tiny cross can be seen etched on the surface of each chamber. (Diagram A)

When the slide is observed under a microscope at a magnification of about x40, the cross appears. Nine large squares are visible, each of which measures 1mm x 1mm. (Diagram B)

Each of the large squares is further divided into either rectangles or squares. For most counting procedures, only the central large square is used. This almost fills the field of view in a microscope when a total magnification of x100 is used. (Diagram C)

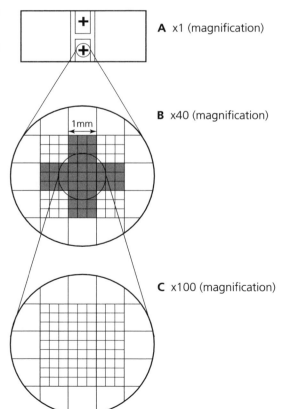

A x1 (magnification)

B x40 (magnification)

1mm

C x100 (magnification)

When using a haemocytometer precautions must be taken to ensure that the count is as accurate as possible:

- Dilution of the suspension

  Most cell suspensions contain too many cells to allow accurate counting, so the original sample must be diluted by a known number of times.

- Filling the slide

  To prepare the chamber it is crucial that the cover glass is positioned correctly. This can be done by breathing on the underside of the cover glass and sliding it horizontally on to the top of the slide, pressing down with the index fingers while pushing with the thumbs:

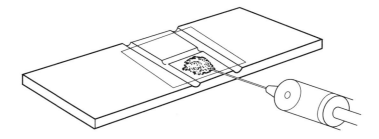

A syringe is used to take a sample of the diluted solution, after it has been well shaken to ensure an even distribution of cells. A portion of the sample is then carefully injected under the cover glass. Only sufficient suspension to cover the silvered front part of the chamber should be used.

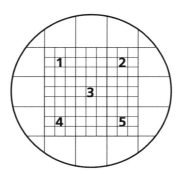

- Counting the cells

  Counting can be carried out on all 25 of the triple-lined squares but this would require a considerable amount of time. Usually, a sample is taken. The simplest method is to count the cells in five of the triple-lined squares.

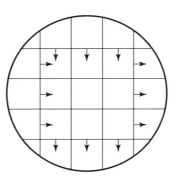

Counting is normally carried out at magnification of x400, when each of the triple-lined squares fills the field of view. For counting it is best if the cells are evenly spread out. Some cells will lie on a boundary of adjoining squares. To ensure that such cells are only counted once the conventional method is to count those cells touching the north and west sides of the square while ignoring those on the south and east.

EXAMPLE:

The central chamber has a 1mm square grid divided into 400 small squares (20 x 20). The side of each small square is therefore 1/20 mm (0.05mm) and the area of each is $1/400\text{mm}^2$.

With the cover glass in place the distance between the central platform and the underside of the cover glass is exactly 0.1mm. Therefore the volume of the suspension above each small square is $1/400 \times 0.1 = 1/4,000\text{mm}^3$ (0.00025mm^3).

A drop of microbial solution is placed on the slide and the number of cells lying above 80 small squares is counted. If 200 cells were counted, these would be suspended in $80/4,000\text{mm}^3$ of liquid. Therefore the number present in 1mm = $200 \times 4,000/80 = 10,000$.

Note, if the sample has been diluted, then the dilution factor must be taken into account.

It's easy to lose track of 'zeros' in this – be careful. 'Guesstimate' an answer so you'll know if your calculation is right or wrong.

Turbidity as a means of counting micro-organisms

A question may ask you to take readings from a calibration curve. Although a small margin of error is allowed, a lot of marks can be lost from careless reading.

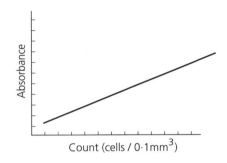

This is known as **colorimetry** and is based on the fact that there is a linear relationship between the number of cells in a culture (estimated by direct counting of samples) and the **optical density** or **turbidity/absorbance** of the culture (measured using a colorimeter). Once a calibration curve has been plotted, the curve of this relationship can be used to convert readings of optical density or turbidity/absorbance to estimates of the number of cells in a sample.

The more cells present, the denser, more turbid will be the suspension.

Fresh mass versus dry mass

Measurements of the **fresh mass** of an organism are relatively quick and easy to make. Fresh mass measurements permit repeated measurements on the same organism. The problem with fresh mass measurements is the likelihood of significant variation in the water content of living organisms, so measurements of fresh mass may give inconsistent readings.

Dry mass measurements of organisms determine the mass obtained after all the water has been removed by drying.

This is more accurate but does require the destruction of the sample in the process.

Working with bacteria

A number of procedures and investigations can be carried out to identify and culture bacteria:

- carrying out an aseptic transfer
- pouring a plate
- preparing a streak plate
- inoculating an agar slope
- gram staining.

Make sure you know the procedures for carrying out these investigations.

Biotechnology and industry

Biotechnology incorporates the way in which plant and animal cells and micro-organisms produce substances which are of use to man.

Making bread and cheese, and brewing beer, were the first examples of biotechnology and date back thousands of years.

Cheese

Pasteurised milk is always used in industrial cheese manufacture, and starter cultures of the bacteria **Streptococcus lactis** and **Strepococcus cremoris** are added.

N.B. The variety of cheese depends largely on the nature of the starter culture, temperature of manufacture and the presence or absence of secondary microbial species on the cheese.

The manufacture of most cheeses involves two major steps:
- curdling of the milk
- ripening of the curd.

Curdling

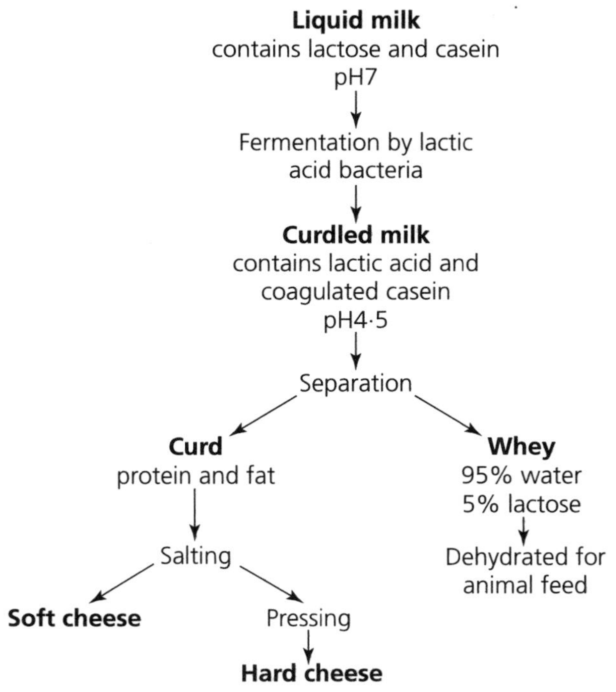

Liquid milk
contains lactose and casein
pH7
↓
Fermentation by lactic
acid bacteria
↓
Curdled milk
contains lactic acid and
coagulated casein
pH4·5
↓
Separation
↙   ↘
Curd   **Whey**
protein and fat   95% water
5% lactose
↓   ↓
Salting   Dehydrated for
↙   ↘   animal feed
Soft cheese   Pressing
↓
Hard cheese

Ripening

This can take between two weeks and 12 months. Lactic acid bacteria or other micro-organisms work on the curd, breaking down milk fat and protein to produce fermentation products, enabling the full flavour of the cheese to develop.

Alcohol

Brewing depends on the fact that yeast cells can live without oxygen, producing carbon dioxide and alcohol from sugar by a process called alcoholic fermentation.

Wine making

- Ripe grapes are picked and crushed to produce a **must** which contains juice, pulp, skins and pips.

- The must is treated with sulphur dioxide to inhibit undesirable bacteria and yeasts on the skins.

- Fermentation or anaerobic respiration is brought about by wine yeast, **Sacchromyces cerevisiae**, which grows on the skins of the fruit.

- As fermentation proceeds alcohol is formed, which extracts the red pigment from the skins of red grapes to produce red wine.

N.B. Rosé wine is produced if the skins are removed before the process is too advanced, and white wine produced if the skins are removed early or if green grapes are used.

- Dry wine is produced if there is a complete breakdown of the sugar, giving an alcohol content of 15% (at which level the yeast is killed).

- Sweet wines are made from musts with a high sugar content or by stopping fermentation before all the sugar is used up.

- Sparkling wines are made by adding sugar to the fermented must, adding wine yeast and bottling the mixture.

You should try to link this to your biochemistry – what exactly is happening biochemically?

Beer making

- This is based on yeast fermentation of an infusion made from grain (usually barley), and compared with wine making is microbiologically a more controlled process.

- Barley grains are germinated under carefully controlled conditions, so that the enzymes (amylases and peptidases) digest the starch in the grain to maltose and the protein to amino acids (**malting**).

- The malting is brought to an end by raising the temperature and after adding warm water this is now called the **wort**.

Don't forget – yeast would normally respire aerobically.

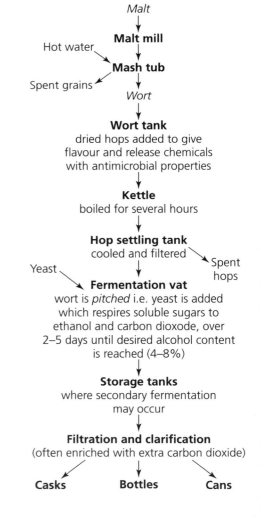

Malt
↓
Malt mill
Hot water ↓
Mash tub
Spent grains ↗ ↓
Wort
↓
Wort tank
dried hops added to give
flavour and release chemicals
with antimicrobial properties
↓
Kettle
boiled for several hours
↓
Hop settling tank
cooled and filtered ↘ Spent
Yeast ↘ ↓ hops
Fermentation vat
wort is *pitched* i.e. yeast is added
which respires soluble sugars to
ethanol and carbon dioxode, over
2–5 days until desired alcohol content
is reached (4–8%)
↓
Storage tanks
where secondary fermentation
may occur
↓
Filtration and clarification
(often enriched with extra carbon dioxide)
↙ ↓ ↘
Casks **Bottles** **Cans**

A note about fermentation

- Fermentation as used in industrial microbiology is often described as a process involving the large-scale culturing of cells in either aerobic or anaerobic conditions.

- The fermentation industry exploits a wide range of microbial processes for the production of commercially useful products and it usually necessitates the large-scale growth of micro-organisms.

- Fermenters, sometimes known as bioreactors, are large reaction chambers which provide a controlled environment for the growth of bacterial and fungal cells.

- Some fermenters have agitation systems to mix air bubbles into the liquid and are used for the aerobic growth of microbes, while others are simple vats for anaerobic growth.

- Fermentation processes may be **batch**, i.e. substrate is added, fermentation occurs and the product is recovered; **fed-batch**, i.e. substrate may be added during fermentation; or **continuous**, i.e. substrate is added and product removed continuously or at intervals.

- The advantage of the continuous process, is that product is removed constantly and substrate added constantly, so the process can be maintained over a long period of time without interruption, resulting in more product on a regular basis.

> Draw up a table to show the advantages and disadvantages of batch and continuous fermentation.

Antibiotic production

- An antibiotic is a chemical substance sometimes produced by micro-organisms which has the ability to inhibit or even destroy other micro-organisms.

- Demand for antibiotics on a large-scale began in World War 2, prompting the need for industrial scale production.

- Antibiotic production is achieved by large scale aerobic growth of selected organisms in **stirred-tank fermenters**.

Tank fermenter

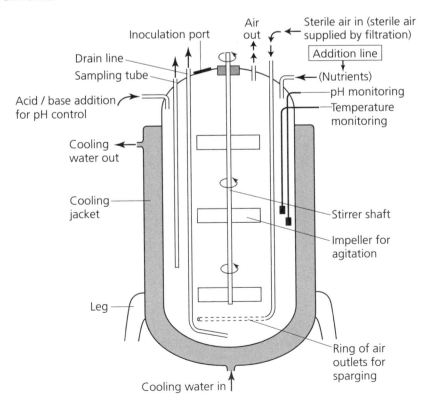

The manufacture of penicillin is typical of antibiotic production:

- Cultures of the bacteria **Penicillium chrysigenum** or **Penicillium notatum** are started in small tanks before being transferred to large fermenters where optimal conditions for cell growth are maintained, including continuous mixing by introducing air at the bottom (known as sparging) and the rotation of impellers.

Penicillin production in batch culture

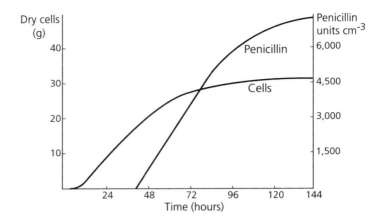

- In the first 24 hours the nutrient (glucose) is in excess and the cells multiply rapidly.

- When the amount of glucose becomes less, growth rate slows (seen by the shallower slope of the cells line).

- Penicillin production is now triggered (an example of a **secondary metabolite**, i.e. not essential for microbial growth and produced towards the end or after the growth phase).

- After 150 hours penicillin production has ceased and the contents of the fermenter are harvested.

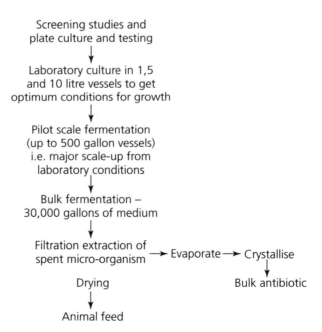

This is a 'scaling up process', i.e. one that starts at laboratory bench level (1, 5 and 10 litre containers) to an industrial level of working with 30,000 gallons of medium.

Downstream processing

The nutrient broth, after fermentation, is a mixture of enzyme, nutrients, waste materials and cells. The enzyme must be extracted and this is done by **downstream processing**.

The aim of downstream processing is to purify as much product as possible at the minimum cost. This depends on the nature of the raw materials, the presence of any undesirable by-products and the concentration of the desired product in the broth.

Stages of downstream processing

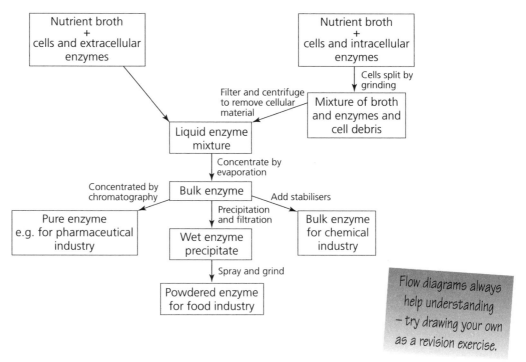

Flow diagrams always help understanding – try drawing your own as a revision exercise.

Immobilising enzymes

Immobilisation is the process of attaching the enzyme to, or trapping it in, an **inert solid support** or **carrier**. This enables the reactants to be passed over the enzyme in a continuous process, or for the enzyme to be used in a batch reactor (a reactor which deals with one batch at a time). The important point is that in both cases the enzyme can be recovered at the end for re-use.

The alternative – mixing the enzyme in solution with the reactants – is technically simpler, but once the reaction is over it involves either wasteful loss of the enzyme or potentially expensive recovery techniques.

Insoluble polymers, in the form of membranes or particles, are typically used to support the enzyme.

Immobilised whole microbial cells are also sometimes used, for example, inside polyacrylanide beads.

Glucose isomerase is currently the enzyme most used in immobilised form and can operate continuously for 1,000 hours at 60°C.

Immobilisation stops enzymes moving around in the substrate and has many uses in industry:

- separation of lactose from whey waste
- microbial production of butanol
- amino acids for food additives
- production of fructose from sucrose, and glucose syrup from starch
- enzyme therapy – reduction of aspargine levels in patients with particular cancers
- diagnostic kits – medical biosensors for detecting glucose/urea levels
- antibiotic production.

Immobilisation can be carried out by a variety of methods.

Chemical bonding

The enzyme is linked to a polymer such as cellulose by a bonding chemical, e.g. glutaraldehyde. This, however, may damage some enzymes.

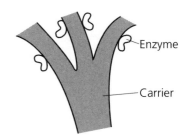

Adsorption

The enzymes or cells are physically adsorbed onto the surface of ion-exchange resins, carbon particles, magnetite, glass beads, ceramic pieces or collagen. The particles may contain an enzyme activator.

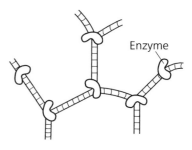

Entrapment

This method is considered to be the most 'gentle' method.

Substrate diffusion can be a problem, and the rate of diffusion depends on the size of the entrapment droplet.

Immobilisation brings advantages to industrial processes.
- It reduces the cost of expensive enzymes.
- The enzymes can be re-used continuously. This reduces industrial 'shut down' time, as well as the overall cost of the enzyme per product.
- An immobilised enzyme may be more easily used in an automated process.
- Does not add to the cost by being difficult to remove from the product.
- May be more effective in a specific reaction because other molecules may be inhibited.

It has been found that immobilisation changes enzyme kinetics, e.g. substrate diffusion, pH tolerance and permitted substrate concentration.

This is because a micro-environment is created round the immobilised enzyme.

Biosensors

- The biosensor uses an immobilised biological molecule, usually an enzyme, to detect or 'sense' a particular substance.

- The biosensor does this by reacting specifically with the substance to be detected to give a product.

- The product of the reaction could be heat, a gas, a soluble chemical or an electric current.

- A number of sensing devices have already been introduced, such as glucose monitors for medical purposes and nerve gas detectors for military use.

- Biosensors are used to find out information about the body without invading the body. For example, the presence of glucose in urine can be detected by dipping a plastic strip coated with the enzymes glucose oxidase and glucose peroxidase and a dye into a sample of urine and noting any colour change. These enzymes are called diagnostic enzymes and there are a number of biosensors which use them:

Substance tested for	Diagnostic enzyme
Alcohol	Alcohol dehydrogenase
Urea	Urease
Penicillin	Penicillinase
Glucose	Glucose oxidase
Cholesterol	Cholesterol oxidase

- Biosensors can be used in industrial processes, e.g. continuous monitoring of an industrial fermentation process to allow conditions such as pH, temperature and substrate concentration to be maintained precisely at optimum levels.

- One major future development in biosensor research is towards making more sensitive miniature sensors called 'biochips'. These would be small enough to implant in the human body. Interfaced with the relevant devices they could, for example, be used to artificially regulate heartbeat or be artificial sense organs.

Genetic engineering

Cell culture

Cell culture may be defined as 'a method of growing cells derived from animals and plants under controlled environmental conditions *in vitro* which includes the culture of single cells'.

In cell cultures the cells are no longer organised into tissues – treatments are used at the beginning of the process to ensure that the tissues are disrupted into individual cells.

The cells can be grown either in an unattached **suspension culture** or attached to a solid surface, e.g. a **monolayer culture**.

Two other culture methods may be used – callus culture and tissue culture.

Animal cell cultures

Animal cells may be grown either in an unattached suspension or attached to a solid surface. Initial growth of cells in culture is termed **primary culture** and these cells cannot be maintained continuously as they die after a few sub-cultures.

N.B. Primary cells will only grow when they are attached to a solid surface.

Occasionally variant cells arise *in vitro* that can be propagated indefinitely as a **cell line**. Most established cell lines may be grown either in monolayer culture or in suspension culture.

Monolayer cultures are started by disrupting a portion of tissue by incubating it with proteolytic enzymes. The enzyme is then removed by washing and centrifugation, and the cell suspension mixed with culture medium in a vessel with a flat surface, so that the cells can settle and grow to form a monolayer. When the culture has covered the vessel surface, the cells are harvested by treatment with buffed enzyme (trypsin) solution. The harvested cells may then be used to prepare fresh monolayers or to inoculate media for suspension solutions.

The development of cell culture techniques has had an impact on medicine and industry:

- propagation of viruses for use in viral studies and vaccine preparation, reducing the need for experimental animals
- synthesis of monoclonal antibodies
- whole organ culture, e.g. the production of sufficient quantities of human skin for grafting after severe burns
- genetic manipulation for synthesising new products and the improved expression of existing ones
- production of metabolites, e.g. interferon for immune cells such as lymphocytes.

Plant cell cultures

If cultures of isolated cells can be maintained, they can be induced to produce specific chemicals normally only produced by mature cells. This means that products formed during different stages of a plant's normal growth and development can be produced in a fermentation type environment, similar to that used for microbial cultures.

The media and physical conditions required for isolated plant cell growth are:

- despite the presence of chlorophyll most cell cultures cannot photosynthesise and so need a carbon source, e.g. sucrose
- nitrogen, usually supplied as ammonia or nitrate
- trace elements supplied as simple salts
- media also contain plant growth regulators such as auxins, gibberellins and cytokinins
- incubation is at 24–27°C and pH ranges from 5 to 7.

N.B. Mass plant cell cultures may need heat to maintain the growth temperature.

Plant cell culture takes place in two stages:

- **fermentation** which uses the culturing methods in industrial microbiology

Batch processing

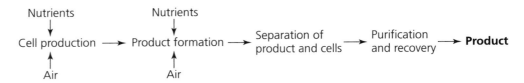

N.B. The response of plant cells to this treatment is different from that of microbes because plant cells are shear-sensitive, i.e. they lyse easily when agitated. The solution to this forms the second stage.

- **immobilisation** of the cells within the suspension culture.

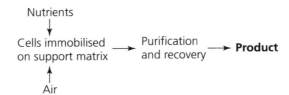

Plant tissue culture

Often called **micropropagation**, this method of plant propagation has several advantages over traditional methods such as seeds, cuttings, etc:

An exam question may ask you to explain each of these – why are they advantageous?

- high rates of genetic uniformity (**clones**)
- large numbers of plants raised in a relatively small space
- aseptic propagation produces bacterial and fungal disease-free plants
- all-year-round production is possible
- plants need little or no attention between subcultures.

Plant growth *in vitro* is either **organised**, i.e. when structures such as shoot or root meristems are cultured and grow and the original structure is preserved, or **unorganised** i.e. growth is characterised by the presence of only a few of the specialised cells found in an intact plant and there is an absence of any recognisable structure.

There are a number of different types of tissue culture:

- **callus cultures** are groups of cells from the unorganised growth of small plant organs, tissues or previously cultured cells

- **suspension cultures** or isolated plant cell cultures are cells or small clumps of cells suspended in an agitated liquid medium

- **protoplast culture** is the process of releasing plant cells without walls (**protoplasts**) from plant tissue and their maintenance in suitable aseptic conditions.

'Organ' culture describes the growth of organised parts of plants such as meristems, shoot tips, embryos, etc.

Tissue culture media comprise a solution of salts, a carbon energy source, vitamins, amino acids, growth regulants. The processes all follow a similar basic series of steps.

Micropropagation

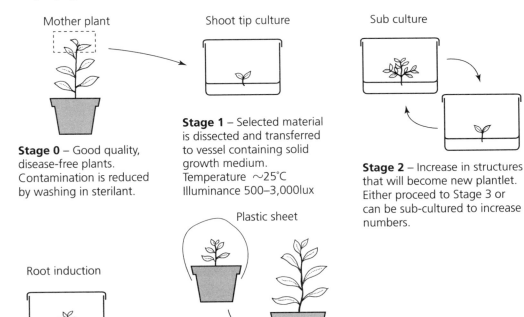

Mother plant

Stage 0 – Good quality, disease-free plants. Contamination is reduced by washing in sterilant.

Shoot tip culture

Stage 1 – Selected material is dissected and transferred to vessel containing solid growth medium. Temperature ~25°C Illuminance 500–3,000lux

Sub culture

Stage 2 – Increase in structures that will become new plantlet. Either proceed to Stage 3 or can be sub-cultured to increase numbers.

Root induction

Stage 3 – Plants grown on to become fully rooted and photosynthetic. Roots induced using auxin.

Plastic sheet

Weaning

Stage 4 – Plantlets weaned from culture conditions so they can withstand more rigorous conditions. Planted in sterile compost and kept in high humidity for several days before gradually being 'hardened off'.

Recombinant DNA technology

In recombinant DNA technology, genes from one organism are introduced into the genome of an unrelated organism to, for instance, 'engineer' new varieties. It is mostly used to produce new varieties of micro-organisms which are then used to manufacture useful chemicals by various biotechnological processes.

Recombinant DNA technology became possible through the use of bacterial enzymes called **restriction enzymes** which cut DNA at specific sites. Some restriction enzymes make a simple cut through both strands of DNA (like scissors) leaving **blunt ends** while others make a staggered cut which leaves **sticky ends**.

'Blunt' and 'sticky' ends

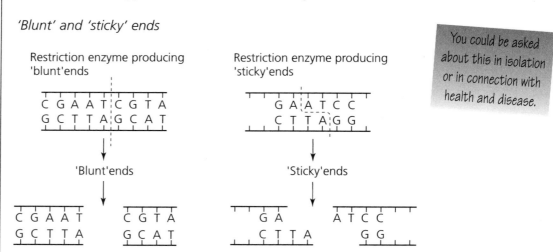

You could be asked about this in isolation or in connection with health and disease.

Any two strands of DNA with complementary bases can be joined together in an enzyme-controlled process known as **annealing**.

Plasmids are extra rings of DNA found in bacteria which are widely used as vectors to introduce additional DNA into host cells. The plasmids can be extracted from bacterial cells, modified by recombinant DNA technology (cut open with restriction enzymes, a new length of DNA added and the plasmid ring annealed) and returned to the bacterium. Here, rapid reproduction produces large quantities of the new genes and/or their products, e.g. the large-scale production of human insulin, interferon and human growth hormone.

There are many complex diagrams explaining this – keep it simple!

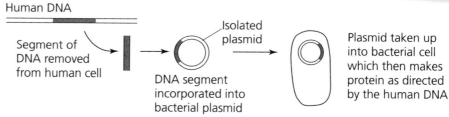

Genetic fingerprinting

- About 90% of eukaryotic DNA does not code for protein production and is described as non-functional.

- These variable and unique lengths of non-functional DNA are passed on to offspring and it is this part of the DNA which is used in DNA fingerprinting.

Genetic fingerprinting

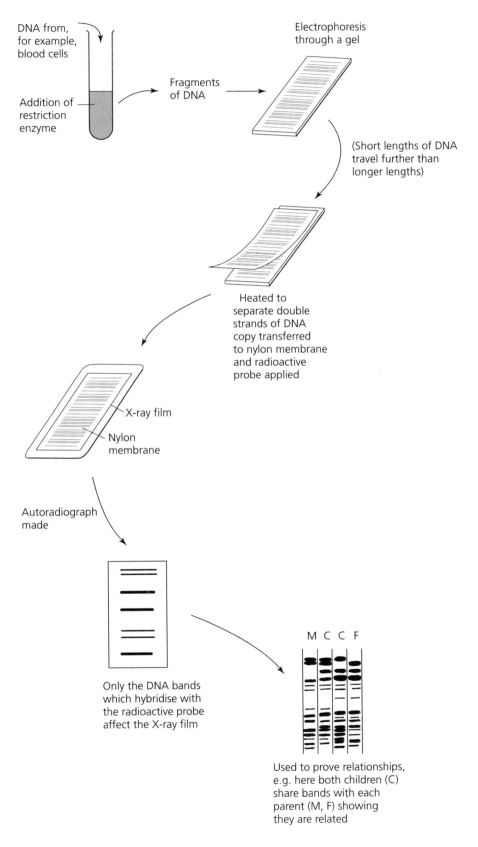

DNA from, for example, blood cells

Addition of restriction enzyme

Fragments of DNA

Electrophoresis through a gel

(Short lengths of DNA travel further than longer lengths)

Heated to separate double strands of DNA copy transferred to nylon membrane and radioactive probe applied

X-ray film

Nylon membrane

Autoradiograph made

Only the DNA bands which hybridise with the radioactive probe affect the X-ray film

M C C F

Used to prove relationships, e.g. here both children (C) share bands with each parent (M, F) showing they are related

Suggested further reading

Beckerstaff, G.F. (1987) *Enzymes in industry and medicine*, Cambridge University Press (0-521-42753-3)

Collett, R.A. & Coles, A. (1993) *Microbiology and technology*, Cambridge University Press (0-521-43687-7)

Fincham, J.P.S. & Ravetz, J.R. (1991) *Genetically altered organisms – benefits and risks*, Open University Press (0-335-09618-2)

Freeland, P.W. (1991) *Micro-organisms in action,* Hodder and Stoughton (0-340-53268-8)

Kirby, L.T. (1990) *DNA Fingerprinting – an introduction*, Macmillan (0-333-54024-7)

Olejnik, I. & Farmer, B. (1989) *Biology and Industry*, Blackie (0-216-92627-0)

Mannion, K. & Hudson, T. (1996) *Microbes, Medicine and Biotechnology*, Collins Educational (0-00-322392-2)

Smith, J.G. (1993) *Biotechnology* (2nd edn), Cambridge University Press (0-521-42784-3)

The Biochemical Society – *Biochemistry across the Curriculum* Number 3 – *Enzymes and their role in biotechnology*

Chapter eight

Human health and disease

Glossary

Benign – slowly growing mass of cells whose effects are limited
Chemotherapy – treatment by chemicals/drugs for a disease
Immunity – the ability of an organism to resist infection
Infectivity – ease with which a bacterium causes infection
Inflammation – reaction to an injury characterised by heat and swelling
Invasiveness – the ease with which a toxin or bacterium spreads in the body
Lysis – bursting of a cell
Malignant – rapidly growing mass of cells which invades other tissues and may destroy them
Opsonisation – preparation of antibody-antigen complex for phagocytosis
Pathogenicity – how toxic a toxin is
Radiotherapy – treatment by X-rays for a disease
Toxin – poison produced by a living organism, especially one formed in the body and stimulating the production of antibodies
Vector – an organism which spreads disease from one individual to another
Vulnerability/risk factors – factors which increase the chance of disease

> Check with your syllabus which diseases you need to know.

A definition for infectious disease – 'an abnormality of an animal caused by a pathogenic organism that affects the performance of the vital functions and usually gives diagnostic symptoms'.

When considering human health and disease, infectious diseases plus other types of disease such as cancer, genetic disorders and nutritional deficiency diseases need to be discussed.

Inductive agents of disease

This includes viruses, bacteria, fungi, protozoans, etc

Viruses

Examples of diseases caused by viruses

> Take the opportunity to link this section with microbiology.

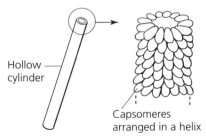

Hollow cylinder

Capsomeres arranged in a helix

Tobacco Mosaic virus

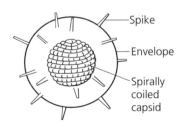

Spike

Envelope

Spirally coiled capsid

Myxovirus

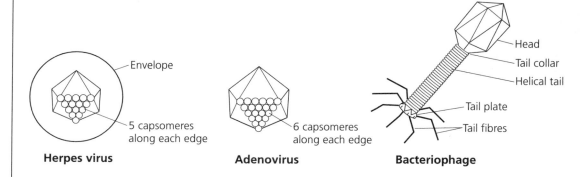

Herpes virus — Envelope, 5 capsomeres along each edge

Adenovirus — 6 capsomeres along each edge

Bacteriophage — Head, Tail collar, Helical tail, Tail plate, Tail fibres

	Naked virions	**Enveloped virions**
DNA viruses	Warts	Chickenpox Herpes simplex
RNA viruses	Poliomyelitis Common cold	Measles Mumps Rabies Influenza Rubella AIDS

Viruses cause disease by disrupting the host cell so that it produces more and more viruses, after which it dies. This bursting is called **lysis** and the mucus that is symptomatic of colds and 'flu is made up of lysed cells. The effects can be more serious, e.g. in **poliomyelitis** the virus damages the cell bodies of the neurones attached to striated muscle. The muscles no longer receive impulses and cannot contract and relax, i.e. they are paralysed.

Bacteria

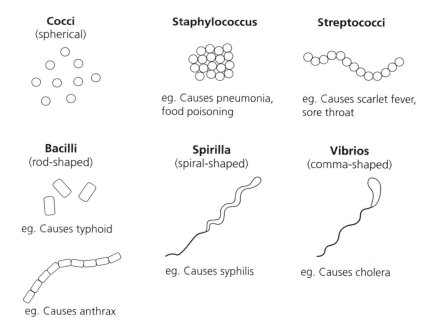

Cocci (spherical)

Staphylococcus
eg. Causes pneumonia, food poisoning

Streptococci
eg. Causes scarlet fever, sore throat

Bacilli (rod-shaped)
eg. Causes typhoid
eg. Causes anthrax

Spirilla (spiral-shaped)
eg. Causes syphilis

Vibrios (comma-shaped)
eg. Causes cholera

Many bacteria cause disease by the production of **toxins**, of which there are two types

- **exotoxins** are secreted during bacterial growth
- **endotoxins** are components of the bacterial cell released when the cell breaks down.

Try to remember the procedure for Gram staining.

The ability of a bacterium to cause disease relies on a number of factors:

- **infectivity**, i.e. the ease with which the bacterium can cause infection
- **pathogenicity**, i.e. the toxicity of the toxin produced
- **invasiveness**, i.e. the ease with which the toxin or bacterium spreads in the body.

Examples of diseases caused by bacteria

Gram positive	Gram negative
Sore throat, rheumatic fever (*Streptococcus pyogenes*)	Typhoid (*Salmonella typhi*)
Pneumonia (*Streptococcus pneumoniae*)	Food poisoning (*Salmonella spp.*)
Boils (*Staphylococcus aureus*)	Cholera (*Vibrio cholerae*)
Tuberculosis (*Mycobacterium tuberculosis*)	Gonorrhoea (*Neisseria gonorrhoeae*)
Diphtheria (*Corynebacterium diphtheriae*)	Whooping cough (*Bordetella pertussis*)
Tetanus (*Clostridium tetani*)	

Fungi

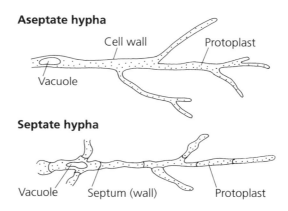

Compared to viruses and bacteria, fungi are of minor importance as animal pathogens.

Fungi mainly cause disease either when large numbers of spores enter a host by chance, or when immunosuppressed individuals are invaded by their normal flora.

Examples of diseases caused by fungi

Clinical type	Examples	Fungi
Superficial infections	Candidiasis of skin, mucous membranes (thrush) and nails	Candida albicans
Deep infections (mainly in immunosuppressed individuals)	Crytococcal meningitis Pulmonary aspergillosis	Crytococcus neoformans Aspergillus fumigatus

Fungi which cause skin diseases are known as dermatophytes. All spore-forming fungi can cause respiratory problems if the spores are inhaled.

Transmission of disease

Only diseases which are caused by **infective agents** can be called **communicable** diseases, others are **non-communicable** diseases.

Mistakes often occur because communicable and non-communicable diseases are confused.

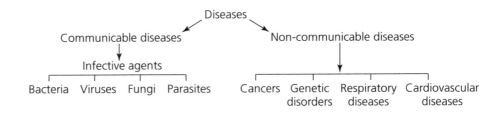

For infection to occur, a minimum 'dose' of micro-organism must enter the body via an appropriate route, e.g. a natural opening (mouth, eyes, nose, etc.) or by wounding, e.g. if an insect vector pierces the skin.

Pathogenic organisms vary in their ability to survive outside the body. The poorest survivors are the sexually-transmitted organisms. The best are the spore-producing bacteria which cause tetanus (*Clostridium tetani*) – its spores survive for years, only germinating if they enter the body through damaged skin.

Disease vectors are mostly blood-feeding insects which take up infected blood when feeding, then re-inject infected saliva into a new host when feeding later. For example, malaria is caused by a small parasitic protozoan transmitted by a mosquito.

Stages of malaria

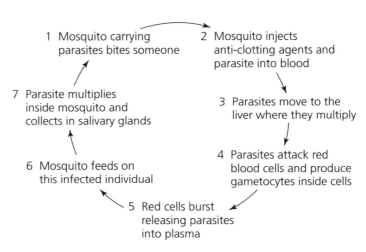

Flow diagrams are always a good way to remember sequences.

1 Mosquito carrying parasites bites someone

2 Mosquito injects anti-clotting agents and parasite into the blood

3 Parasites move to the liver where they multiply

4 Parasites attack red blood cells and produce gametocytes inside cells

5 Red cells burst releasing parasites into plasma

6 Mosquito feeds on this infected individual

7 Parasite multiplies inside mosquito and collects in salivary glands

The body's reaction to infection

It is the body's immune system which restricts access of microbes to the body or destroys microbes should they enter the body.

The immune system comprises a variety of molecules and cells whose activities often interact to exert an effect.

As a revision exercise redraw this flow diagram with named examples.

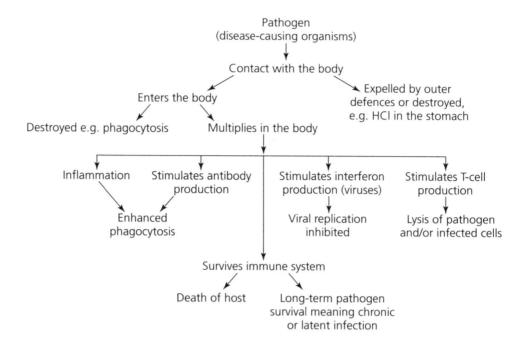

Outer defences

Most outer body defences are non-specific in that they are not designed to combat any particular organism.

Redraw the table with named examples.

Defence Site	Defence Mechanism
Skin	Physical barrier
Respiratory tract	Mucus traps microbes and cilia, carries mucus away from lungs
Stomach	Hydrochloric acid
Large intestine	Commensal bacteria compete with parasites
Secretions like tears	Antibodies and lysozyme (degrade cell walls)

Inner defences

The inner defences of the body differ in that although some are also non-specific, there are others which are specific, e.g. T-cells (see later).

Inflammation is a reaction to injury and is characterised by heat, swelling, redness and pain caused by an increased blood flow. Inflammation helps to direct cells and molecules of the **immune system** to sites of infection by 3 mechanisms:

* increased blood supply means an increased supply of antibodies, neutrophils and lymphocytes
* increased capillary permeability allows soluble substances such as antibodies to reach the site of infection
* attraction of phagocytes to the infected site.

Inflammation

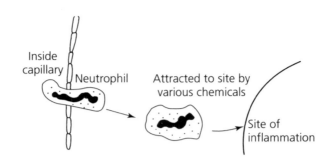

On the surface of a cell membrane are proteins and glycoproteins which are known as **antigens**.

Think about other occasions where surface recognition occurs.

The more closely related organisms are, the more antigens they are likely to have in common.

Lymphocytes recognise the presence of a foreign antigen – these are white blood cells made in the white bone marrow and each lymphocyte recognises *one* specific antigen.

There are two types of lymphocyte:

- **T-cells** which are activated by the thymus gland and are involved in the **cell-mediated** response
- **B-cells** which are activated by cells of the bone marrow, spleen and liver and are involved in the **humoral response**.

The cell-mediated response

T-cells have receptor proteins attached to their cell membranes. When a T-cell which has arrived at an infection site comes into contact with a **complimentary antigen**, i.e. an antigen it recognises on another cell, the antigen is bound to the receptor proteins and the cell is destroyed.

This reaction results in the T-cell undergoing a rapid series of cell divisions, producing **clones** of identical T-cells which are all capable of recognising and destroying the same type of antigen.

Another name for these cells is **killer lymphocytes** or **killer cells**, because of the way they destroy invading material.

The humoral response

B-cells, like T-cells, have receptor proteins on the cell membrane surface, but when they detect a complimentary antigen and bind to it, the effect is different.

Take the opportunity to revise the structure of the blood.

B-cells undergo rapid divisions and two different types of cell result – **plasma cell clones** and **memory cells**.

Most of the new cells are plasma cell clones and these have the ability to produce large quantities of **antibody** (special proteins released into the circulation which bind to the antigen causing its destruction).

In other words, the B-cells and plasma cell clones do not directly destroy the antigen.

This is because they live for only a few days, but in that time they produce up to 2,000 antibody molecules per second!

The other product of B-cell division, memory cells, are concerned with rapid response should a second invasion of the antigen occur.

This is known as an **immunological memory** and is conveyed by memory cells which cause the rapid production of plasma clone cells should the antigen be encountered.

Response to antigen

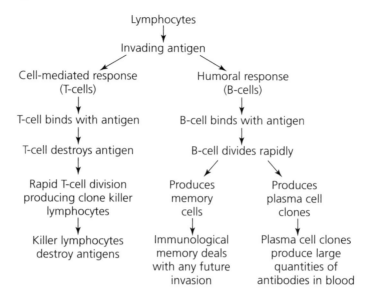

How antibodies work

Antibodies are large protein molecules with attached carbohydrates (glycoproteins), and are made from two types of polypeptide chains, heavy and light, joined by disulphide bridges.

Antibody structure

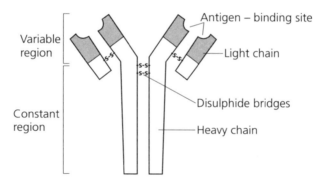

The variable portion of the polypeptide chains results in many different antibodies, each showing specificity for a different antigen.

Antibodies are known collectively as **immunoglobulins** and they contribute to the body's defences in a number ways:

- antibodies reduce the ability of most pathogens to invade host cells

- they bind to antigens and then clump together or **agglutinate**, preventing spread of the antigens

- they have two active sites so each antibody can combine with two antigens – if the antigens are 'shared' between the active sites of two antibodies a structure called an **antibody-antigen complex** is formed

- **phagocytes** more readily engulf antibody-antigen complexes (this 'preparation' for phagocytosis is called **opsonisation**).

A number of different types of immunoglobulins have been identified, each having a slightly different function:

- **IgM** is first to appear when an antigen is detected, is produced for 1–2 weeks and concerned with the body's **primary response** (see later)

You don't necessarily need to know named examples of immunoglobulins but you must know how they work.

- **IgG** makes up 80% of immunoglobulin and is concerned with the body's **secondary response** (see later)

  N.B. IgG cannot be made during the first months of life, so a baby is dependent on its mother's immunoglobulins found in breast milk and passed across the placenta

- **IgA** is found at low levels in body secretions such as tears or colostrum in breast milk

- **IgD** and **IgE** are found in very small amounts and their functions are uncertain, but IgE may be concerned with allergic reactions.

The immune response

The **primary response** occurs 3–14 days after infection of the body by an antigen for the first time.

The period between infection and the start of antibody production is called the **latent period**, after which the level of antibody in the blood rises rapidly before falling again.

Memory cells are produced during this time and remain in the body ready to attack a second invasion of antigen.

The **secondary response** is what happens if a second infection occurs and it is triggered by a smaller amount of antigen. This response is more rapid and more antibody is produced than in the primary response, which means that the pathogen is destroyed before the body is infected and symptoms are detected.

This is why most people suffer from chickenpox only once (there is only one strain of chickenpox virus) but will suffer from the common cold many times (there are many antigenic forms of the common cold virus).

Primary and secondary immune responses

> Being able to remember and explain this graph will produce a good answer to a question about immune response.

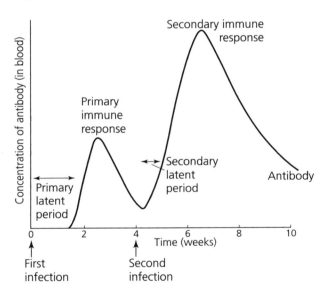

Protection against infection

Specific immunity can be acquired either **actively,** when a person's own immune system responds, or **passively**, when the immune system of another person responds and antibodies are transferred.

Active and passive immunity can be acquired through natural and artificial means:

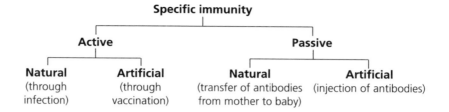

Natural immunity

When a body comes into contact with an invading antigen, antibodies are formed to destroy the antigen. This is known as **natural active immunity**.

There is another form of natural immunity and it is that provided by a mother via the placenta and breast milk; it is temporary and lasts until the baby's own immune system is active. This is known as **natural passive immunity**.

Artificial or induced immunity

As with natural immunity, there are two forms.

- When antibodies are formed in one individual, extracted and injected into another, this is called **acquired passive immunity**. This does not confer long-term immunity, but is used if an individual is suspected of having been exposed to an antigen.

- In **acquired active immunity**, small amounts of antigen are given to induce immunity in an individual. The antigen is made safe (see later) and the body produces antibodies and forms appropriate memory cells. If the active antigen is later encountered, the body will destroy it and the disease will not be experienced.

Vaccines

A vaccine is a preparation containing antigens which, when administered, will stimulate active immunity and protect against infection.

Vaccines can be made in a number of ways:

- the virulent micro-organism is cultivated, harvested and then killed; this is known as a **dead or inactivated vaccine**, e.g. whooping cough

- an avirulent but live vaccine, often a mutated form that does not cause the disease, is called an **attenuated vaccine**, e.g. polio, measles, rubella

- when the toxin produced by the pathogen causes the disease, this toxin can be chemically modified so that it still resembles the original but can no longer cause the disease; this is known as a **toxoid vaccine**, e.g. tetanus, diphtheria

- antigens are extracted from the micro-organism and used as a vaccine, e.g. glycoproteins purified from an influenza virus

- cloning of the gene for the relevant antigen in bacterial, yeast or animal cells is carried out on a large scale, e.g. hepatitis B.

N.B. Immunisation is often used synonymously with vaccination but it has a slightly broader definition since it can also include the injection of antibodies, i.e. the induction of artificial passive immunity.

Chemical treatment of infection

This is the treatment of an infection with a chemical which harms the infecting microbes without causing serious harm to the individual.

- Many **chemotherapeutic agents** are available for treating fungal and bacterial infections. The range for treating viral infections is smaller, because it is difficult to attack a virus without harming the cell in which it is replicating. A few anti-viral drugs are available, e.g. zidovudine used in AIDS treatment.

- **Synthetic agents** such as the sulphonamides are available. These have a basic ring structure resembling the para-aminobenzoic acid (PABA) growth factor that many bacteria need to make folic acid. Since the bacteria cannot distinguish the sulphonamide from PABA they make a non-functional molecule instead of folic acid, e.g in the treatment of urinary tract infections.

Blood groups

Human blood exhibits a specialised aspect of the immune response.

The surfaces of red blood cells carry molecules which act as antigens – they are **mucopolysaccharides** known as **agglutinogens**. There are two different ones, **A** and **B**.

In the plasma there are two antibodies called **agglutinins**, **a** and **b** (these are not made in response to the presence of antigens but are present anyway).

If red blood cells carrying a particular agglutinogen come into contact with plasma containing the complementary agglutinin, a reaction occurs which causes the red blood cells to stick together or **agglutinate**. This means the red blood cells cannot carry out their normal function – large scale agglutination is fatal.

Blood group	Agglutinogen present	Agglutinin present
A	A	b
B	B	a
AB	A and B	none
O	none	a and b

Most of the time blood groups are not important, but they become vital if a transfusion is needed. During a transfusion it is the cells of the donated blood that can be affected by an adverse reaction (there is usually much more recipient's blood so any adverse reaction involving it is small).

Blood group O can be given to anyone in a transfusion, because it carries no antigens to stimulate agglutination, and is known as a **universal donor**.

AB is known as a **universal recipient**, because the plasma contains no antibodies to cause agglutination.

ABO Compatibility

		Recipient blood			
		A	B	AB	O
Donor blood					
	A	✓	x	✓	x
	B	x	✓	✓	x
	AB	x	x	✓	x
	O	✓	✓	✓	✓

✓ no agglutination/transfusion OK

x agglutination/transfusion not OK

Drawing a table of transfusions always helps get your thoughts in order.

The **rhesus factor** is an agglutinogen found on the red blood cells of some people who are subsequently known as Rhesus positive (Rh+)

N.B. A Rhesus negative (Rh−) person normally has neither antigen nor an anti-rhesus antibody.

The anti-rhesus antibody is only produced if Rh+ blood is introduced into the bloodstream of a Rh− person.

The first time this happens it does not cause a particular problem. If it happens a second time there is a rapid reaction which causes agglutination, e.g. if a Rh− mother carries an Rh+ baby she may develop agglutinins in her plasma to any foetal red cells which may leak back across the placenta. Subsequent pregnancies with Rh+ foetuses may result in the mother's agglutinins causing the foetal blood to agglutinate – a condition known as **haemolytic disease** of the new-born.

> This is a popular question. Remember it's always the second pregnancy that poses the problem.

Transplants and the immune system

The development of the transplantation of organs and tissues from one person to another has led to much work being done on the suppression of the immune system. Transplanted organs can easily be rejected by the body's defence mechanism because the immune system recognises the foreign antigens or transplant.

In transplantation the most important antigens are the **human leucocyte antigens**. HLA proteins are determined by 5 different genes, each of which has a lot of alleles. Tissues are **compatible** if the HLA proteins match. If they are not compatible **rejection** will take place as the recipient's T-cells and antibodies attack the tissue. Only identical twins have the same HLA proteins. Before a transplant can take place **tissue typing** is done to find the closest possible HLA match between potential donor and recipient.

Even when the match is close, there is still a possibility of rejection therefore treatment after transplant involves ways of suppressing the immune system so that the T-cells do not recognise the transplanted tissue as foreign:

- **chemotherapy** is the use of drugs to suppress the immune system – these are known as immunosuppressant drugs, e.g. cyclosporin A which only affects T-cells and leaves the B-cells to resist infection

- **radiotherapy** is the use of X-rays to irradiate bone marrow and other areas which produce blood, to inhibit leucocyte production.

Examples of diseases

Cancer

This is any malignant growth or tumour resulting from an abnormal or uncontrolled division of body cells.

Cancer is one of the most common diseases of developed countries, causing roughly 20% of all deaths.

Lung cancer is the most common form in men causing 40% of cancer deaths, while 25% of cancer deaths in women are from breast cancer.

> Questions about cancer are often linked to cell division. Take the opportunity to revise mitosis and meiosis.

There are over 100 different forms of cancer with one factor in common – they are the result of uncontrolled mitosis. In a cancerous cell something goes wrong with the control of the cell cycle and the cell divides more than it should, with the resulting mass of cells forming a **tumour**.

A **benign tumour** is a slowly growing mass of **non-cancerous cells** whose effects are limited.

A **malignant tumour** is a rapidly growing mass of cells which invade other tissues and may destroy them. These malignant cells typically get into the circulation and set up secondary 'colonies' elsewhere in the body.

Tumour growth

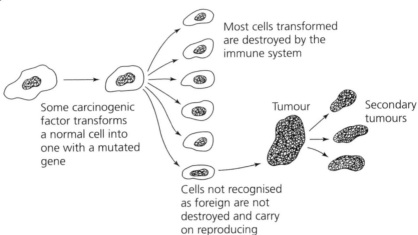

Some carcinogenic factor transforms a normal cell into one with a mutated gene

Most cells transformed are destroyed by the immune system

Cells not recognised as foreign are not destroyed and carry on reproducing

Tumour

Secondary tumours

> Remember the different types of mutation...

A **mutation** in a gene that controls cell division results in the lack of control of cell division. The causes of these mutations are many and the causal agents are called **mutagens**:

- a number of **industrial chemicals** are known to cause cancer and are described as being **carcinogenic**, e.g. asbestos

- **cigarette smoke** contains many different compounds, but it is the **tars** that are carcinogenic

- a **low fibre** intake in the diet increases the chances of getting bowel cancer, and it is known that cancers of the breast and bowel are common when **fat levels** in the diet are high, although the details are still to be researched

- several sorts of ionising **radiation** may lead to cancer, e.g. excessive **UV radiation** may lead to skin cancer; **X-rays** can damage genetic material resulting in cancer

- **nuclear power stations** and **coal mines** produce small amounts of radiation and there is some evidence that the children of nuclear power station workers have a higher risk of developing leukaemia

- some cancers are known to be caused by **viruses**, e.g. cervical cancer is associated with human papilloma virus

- cancer can be more common in some families than others, suggesting a **genetic link**.

N.B. It is *not* the disease itself that is inherited but a susceptibility to factors which cause the disease.

One treatment for cancer is **surgery** but it is not always appropriate, especially if the cancer has spread. A second treatment is the use of high energy ionising radiation to destroy the cells – **radiotherapy**. A third approach is the use of cytotoxic drugs, which suppress cell division, i.e. **chemotherapy**.

Retroviruses are thought to cause a number of cancers, e.g it is thought the **Herpes simplex** virus, which can be transmitted during sexual intercourse, may be responsible for cervical cancer.

RNA — Protein spikes
— Reverse transcriptase
— Capsid
— Envelope

On entry into host cell the envelope is lost ↓

— Reverse transcriptase
— Viral RNA

— Viral RNA
— New viral DNA produced by transverse transcriptase

Viral DNA inserted into host DNA

If the new gene acts as an 'oncogene' the cell becomes a cancer cell

Transcription produces many RNA molecules

Respiratory diseases

Smoking is a major cause of respiratory disease, although other factors also contribute, such as pollution and some occupations where certain particles can be inhaled.

Bronchitis

Someone is said to have **chronic** bronchitis if they have a cough which lasts for at least three months over two years with no apparent cause. Another symptom of bronchitis is that sufferers get out of breath easily.

Smoking and air pollution paralyse the cilia in the bronchial tubes. This prevents them from beating rhythmically, so mucus accumulates and has to be coughed up. The bronchial lining becomes inflamed and irritated and without the mucus to trap irritants and micro-organisms, further irritation and inflammation can develop.

Smoking also damages the walls of the bronchioles and alveoli, causing fibrous growth which narrows the airways.

Emphysema

The walls of the alveoli are broken down in emphysema, resulting in a reduced surface area for gas exchange. Emphysema is most common in middle-aged men who smoke, although it is also associated with occupations such as coal mining.

Bronchiole

Healthy alveolus with large surface area for gas exchange

Unhealthy alveolus with damaged walls and reduced surface area

Questions on this topic are often linked to questions about the structure of the system...

Cardiovascular disease

- Diseases of the blood vessels are responsible for a very large number of deaths. Most are either due to **atherosclerosis** (an arterial disease caused by the build up of fatty deposits, known as **atheroma**, on the inner lining) or the progressive degeneration of artery walls. Atheroma in the cardiac arteries can lead to the acute chest pain associated with **angina**, because the cardiac muscle no longer has an adequate blood supply.

Development of atherosclerosis

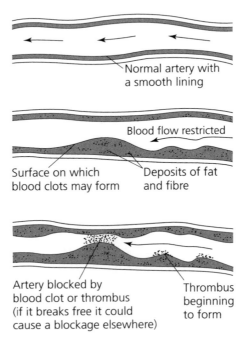

Atherosclerosis can lead to raised blood pressure and **aneurysm**, which is a weakening of an artery wall which can burst at any time.

A blood clot formed at an atherosclerotic site may break away and block a small artery forming an **embolism**. If the blockage occurs in one of the coronary arteries and is left untreated it can lead to a **myocardial infarction** or **heart attack**.

Coronary arteries damaged by atherosclerosis can be surgically by-passed by parts of veins or arteries taken from other parts of the body.

If a thrombosis or embolism blocks an artery supplying the brain the consequent lack of oxygen results in a **stroke**.

Vulnerability/risk factors include:
- age: increases risk but condition can start early in life
- diet : high levels of cholesterol in the diet are associated with a high risk of cardiovascular disease
- weight : obesity increases the stress on the heart
- heredity : heart disease can be more common in some families
- smoking : heavy smokers are much more likely to develop atherosclerosis
- exercise : the physically inactive are more at risk from heart disease.

Diet and exercise

About 2/3 of the people in the world do not have enough to eat and suffer from **malnutrition**. These people rarely die from under-nutrition but are more likely to die from a disease such as measles, because their immune system is vulnerable.
Malnutrition tends to be associated with a lack of food but it

Don't fall into the trap of thinking of malnutrition as only being linked with developing countries.

can equally be applied equally to an excess of food resulting in **obesity**. Obese people die younger, are vulnerable to diseases such as coronary heart disease and diabetes, and suffer from hypertension.

Persistent over-consumption of **alcohol** increases the risk of harm to every bodily function:

- short term effects on the brain
- long term brain and nerve damage
- increased risk of oesophageal and pancreatic cancer
- obesity
- damage to the immune system, resulting in greater susceptibility to infection
- high blood pressure
- weaker and less regular heartbeat
- inflammation of the stomach lining
- liver damage including cirrhosis.

Cholesterol is found in association with lipoproteins, which have an important role to play in the transport of insoluble fats in the blood.

Lipoproteins can be divided according to their density (the less dense lipoproteins contain a higher proportion of lipid to protein):

- **very low density lipoproteins (VLDL)** are formed in the liver and transport triacylglycerol to the tissues
- **low density lipoproteins (LDL)** are rich in proteins and cholesterol and are made from VLDL in the tissues
- **high density lipoproteins (HDL)** are the form in which excess `cholesterol is transported to the liver.

How is this linked with cardiovascular disease?

Excess LDL leads to cholesterol deposits in the walls of arteries. LDL cholesterol can be reduced by:

- reduction in excess body fat
- reduction of fat, especially foods rich in saturated fats, in the diet.

When less fat has to be transported about the body, less cholesterol is produced by the liver.

Some body fat is essential to health. Some people mistakenly believe this is still too much and weight is reduced either by extreme dieting (**anorexia nervosa**) or by binging and vomiting (**anorexia bulimia**). So much weight may be lost that the individual dies from starvation.

Regular, sustained exercise brings about modifications in the body which can have long-term benefits to overall health:

How is heart rate controlled?

- the response of the **heart** to exercise is to increase its rate of beating – as a result of increased exercise, the heart muscle becomes stronger and more efficient, and the resting heart rate drops
- regular exercise has been shown to increase blood levels of HDLs and remove cholesterol from the blood
- most studies show that regular exercise results in increased lung volume and efficiency, and an increase in the capillary blood supply to the alveoli
- plenty of exercise tends to increase overall bone mass and keeps joints flexible
- when muscles are used regularly they become larger and stronger, which has the effect of 'toning' the body.

A programme of regular exercise has benefits not only for the fitness of the cardiovascular system, but also in the loss of weight and the maintenance of weight loss.

Genetic disease

About 2,300 genetic diseases are known or thought to result from mutant alleles of normal genes. They can be divided into four groups:

- autosomal dominant
- autosomal recessive
- sex-linked
- chromosomal defects.

Autosomal dominant inheritance

E.g. *Huntington's chorea*
Cause: autosomal dominant gene on the short arm of chromosome number 4
Frequency: 5 per 100,000 and can be heterozygous or homozygous for the defective gene
Symptoms: onset in 40s to 50s, involuntary muscle movement and progressive mental deterioration
Treatment: none known
Detection: use of DNA probes can detect carriers of the gene

Take the opportunity to revise genetic terms and the different types of inheritance.

Autosomal recessive inheritance

E.g. 1 *Cystic fibrosis*
Cause: autosomal recessive gene on the long arm of chromosome number 7
Frequency: 1 in 1,600 in Britain
Symptoms: disorder of mucus-secreting glands especially in pancreas, intestines and lungs, resulting in very thick, sticky mucus which can block ducts
Treatment: regular physiotherapy and treatment with antibiotics
Detection: use of DNA probes to find carriers and chorionic villus sampling for detecting sufferers pre-natally

E.g. 2 *Sickle cell anaemia*
Cause: point mutation bringing about change in structure of beta-chain of haemoglobin
Symptoms: cells containing abnormal haemoglobin become sickle-shaped at low oxygen levels and block vessels
Detection: prenatal diagnosis using DNA probes and the heterozygous condition (sickle-cell trait) can be detected using electrophoresis of haemoglobin

Sex-linked inheritance

N.B. Most sex-linked traits are recessive and since the mutation is on the X chromosome, most sufferers are male.

E.g. 1 *Haemophilia*
Cause: X-linked recessive mutation in the gene coding for clotting factor VIII
Frequency: about 1 in 10,000 males
Symptoms: excessive bleeding both internally and externally, usually following injury
Treatment: some forms can be treated with regular doses of factor VIII
Detection: 85% of female carriers can be detected by assaying the level of factor VIII in their plasma

E.g. 2 *Duchenne muscular dystrophy*
Cause: X-linked recessive
Frequency: 1 in 5,000 males
Symptoms: in young boys difficulty in walking, climbing stairs, etc., becoming confined to a wheelchair by the age of 10 years and dying before 20
Treatment: none available
Detection: DNA probes detect carriers in almost all families

Chromosomal defects

Abnormalities in chromosomes are usually so serious that those affected are often unable to reproduce.

E.g. *Down's syndrome*
Cause: trisomy 21
Frequency: 1 in 700 births but can be as high as 1 in 50 depending on maternal age
Symptoms: characteristic appearance with learning difficulties varying from very mild to severe, short life expectancy
Treatment: antibiotics and heart surgery can increase life span
Detection: pre-natally by amniocentesis or chorionic villus sampling

Other examples of chromosomal defects involving the sex chromosomes include:
- Turner's syndrome – XO
- Klinefelter's syndrome – XXY
- Triple X – XXX
- XYY syndrome
- Lethal – YO.

Genetic screening

This is the systematic testing of foetuses, new-born children or individuals of any age to ascertain potential genetic handicaps in either them or their offspring. Pre-natal testing can be in the form of **amniocentesis**, which involves testing a sample of amniotic fluid.

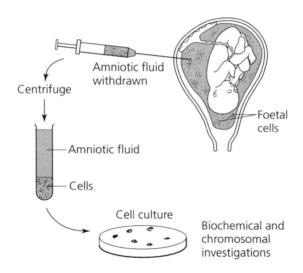

Chorionic villus sampling involves removing some cells from the villi of the placenta for testing

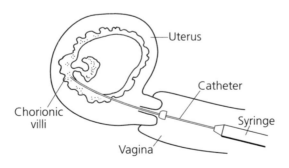

The development of recombinant DNA technology means that in the next few years many more genetic diseases will be detectable pre-natally. A gene probe can be used to detect a fault. This consists of a length of DNA with a specific nucleotide sequence which is used to seek out complementary sequences in the DNA being tested, i.e. in a possible sufferer from, or carrier of, a genetic disorder.

Suggested further reading

Bradley, A. (ed.) (1991) *Healthy Eating – fact or fiction*, Hodder and Stoughton (0-340-488751)

British Medical Association (1991) *Pesticides, Chemicals and Health*, Hodder and Stoughton (0-340-54924-6)

Taylor, D. (1993) *Human Physical Health*, Cambridge University Press (0-521-31306-6)

The Biochemical Society – *Biochemistry across the curriculum* Number 5 – *Immunology*

Walker, A. (1992) *Human Nutrition*, Cambridge University Press (0-521-31139-X)

These pages are for your own notes.

Index